I0900995

IMAGES
of America

MILLINGTON

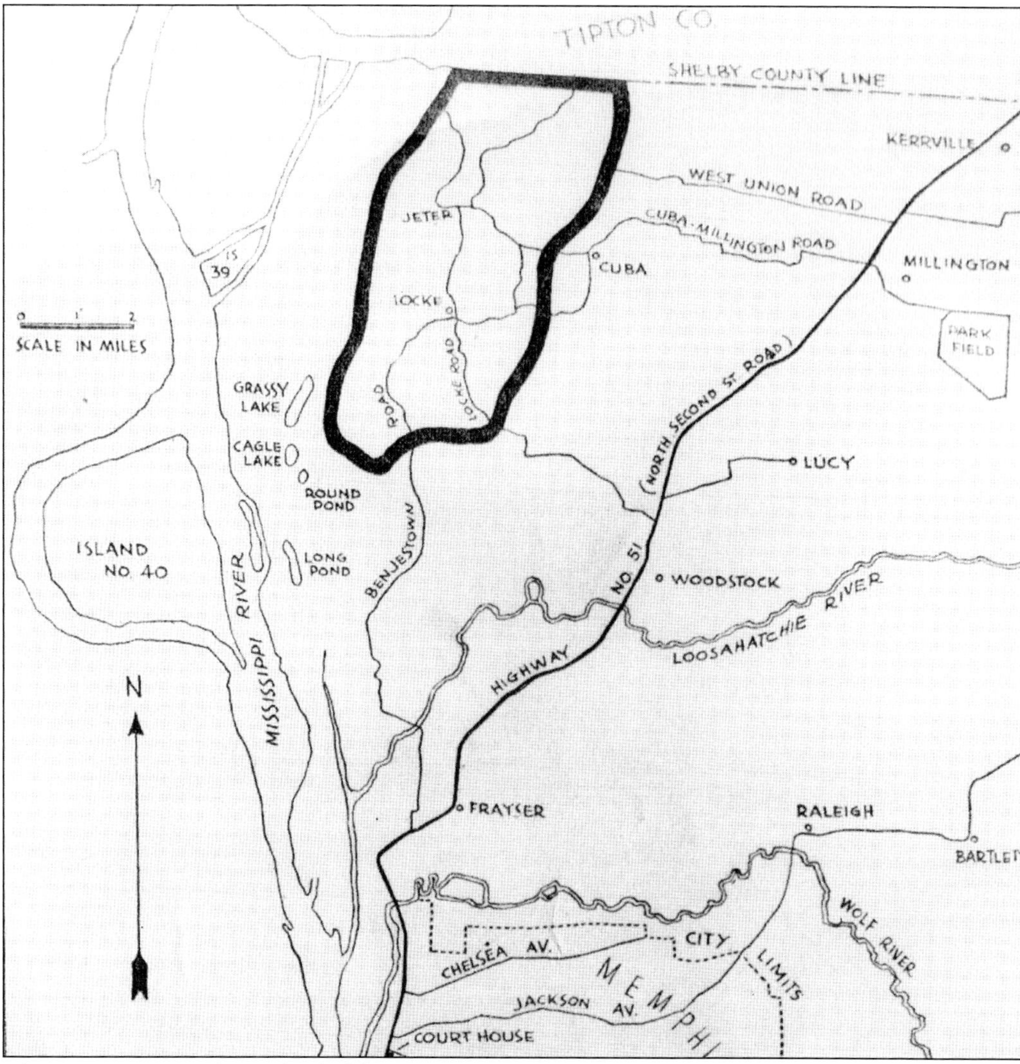

This 1934 map outlines (in bold) the proposed Shelby Forest Demonstration Area west of Millington. Now known as Meeman-Shelby Forest State Park, the park actually extends to the Mississippi River on the west. Many residents of communities north of the Loosahatchie River from the Mississippi River to several miles east of Millington have long enjoyed Shelby Forest and have found doing business in Millington more convenient than traveling to Memphis. (Courtesy the University of Memphis, Mississippi Valley Collection.)

On the Cover: Faye Ellis Osteen reported in her centennial history of Millington that the town's volunteer fire department was established in 1949 with a 10-year-old surplus fire engine. Before then, volunteers responded to fire emergencies but not as an organized and trained unit. James "Buddy" Oakley, pictured, was a trained volunteer. (Courtesy the University of Memphis, Mississippi Valley Collection.)

IMAGES
of America

MILLINGTON

Rita Hiltenbrand Hall

ARCADIA
PUBLISHING

ISBN 978-0-7385-9128-5

Published by Arcadia Publishing
Charleston, South Carolina

Printed in the United States of America

Library of Congress Control Number: 2011937509

For all general information, please contact Arcadia Publishing:
Telephone 843-853-2070
Fax 843-853-0044
E-mail sales@arcadiapublishing.com
For customer service and orders:
Toll-Free 1-888-313-2665

Visit us on the Internet at www.arcadiapublishing.com

This book is dedicated to the pioneers, sojourners, newcomers, and citizens who have forged the character of Millington and its surroundings.

Contents

Acknowledgments

Very little has been written about the history of Millington, and I wanted to help change that. I would like to thank Faye Ellis Osteen (author of *Millington, the First Hundred Years*) and Joseph Marcello for their insights into the past and present of Millington, which contributed greatly to the decision to prepare this book. I am also indebted to Mr. Marcello and Teresa Bolka at Old Timers Restaurant for putting me in touch with others who helped further the cause. I would like to extend special thanks to Laura Rosas at Millington Telephone Company for her generosity with both images and information about growing up in Millington. Ed Frank and the staff of the Mississippi Valley Collection at the University of Memphis library have been especially helpful, as have Vincent Clark at the Shelby County Archives and Sarah Frierson and Wayne Dowdy in the Memphis Room at the Benjamin L. Hooks branch of the Memphis Public Library. I would also like to thank Thomas Sellers at the *Millington Star*, Brian Miller at the Millington Library, Doug Ammons at the Shelby Forest General Store, and my cousins Gwen Andrews, James Clifton, and Jamie Scholik for allowing me access to decades of old photographs, news clippings, and records that have helped me better understand the history of the Millington area and its people. In addition, I am grateful for the guidance and moral support provided by Dr. Janann Sherman and Dr. Beverly Bond at the University of Memphis, Robert W. Dye, and my editor at Arcadia Publishing, Elizabeth Bray. This book would not have been possible without all of you—nor without the patience and support so graciously given by my husband, my son, and my mother, who all now know far more about Millington than they ever cared to. Unless otherwise noted, all images appear courtesy of the University of Memphis, Mississippi Valley Collection. The Millington Telephone Company's Old Timers Restaurant photograph collection (MTC), the Memphis Public Library's Memphis Room (MPL), and the Shelby County Archives (SCA) were other significant sources and will be credited by the foregoing abbreviations.

Introduction

The June 3, 1964, edition of the *Memphis Press-Scimitar* newspaper boldly proclaimed, "Memphis Navy Made Millington a City." Citizens of Millington might not completely agree, however. While the Navy has certainly influenced the population size and the local economy, Millington was well underway before the arrival of the military and continues to thrive today, with the military presence greatly reduced from prior years. Since World War I, members of the Army Air Forces, Navy, and Marines have rotated through Millington, and though some have returned in their retirement years, few have stayed longer than a few years. It is the descendants of 19th-century settlers, the abiding newcomers throughout the 20th century, and the residents of today who have made and continue to make Millington a great place to "Come Home" to, as the town slogan suggests.

The city of Millington, Tennessee, was incorporated in 1903, but like every town, it has a history much older than its charter would seem to indicate. The first settlers to this portion of north Shelby County arrived within the first two decades of the 19th century. Col. Clement McDaniel had a well-established homestead and an active trade going with the local Native Americans when surveyor John Ralston, quite to his surprise, stumbled upon the McDaniels while plotting land claims for North Carolina's Revolutionary War veterans sometime before 1820. In 1822, John Ralston married the colonel's daughter Lucy and settled near the McDaniel homestead. It is rumored that the community of Lucy is named for this young bride, and the Ralstons came to be great landholders in the area.

Nearby, a string-type settlement called Big Creek was forming along the banks of the tributary of the Loosahatchie River for which it was named. Other area communities were named for their prominent citizens, such as Kerrville, named for Rev. Andrew Kerr, and Jeter, named for a pioneer family. Later, communities were given names whose origins, despite their often romantic intrigue, are now forgotten or obscured, such as Glencoe, Millwood, Roaring Falls, Woodstock, Shakerag, Cuba, Locke, West Union, and Independence. Over time, each of these, for whatever reason, was overshadowed by the growing Millington community.

Many of the earliest settlers of this region were wealthy planters who took advantage of the rich soil to grow cotton, the South's staple crop, and to make fortunes in the hardwood industry. Clearing trees to create more arable land resulted in widespread erosion and unpredictable, but frequent, flooding as Big Creek climbed out of its bed to ruin crops and destroy homes. Over time, small villages such as Glencoe and Millwood were established at seemingly safe distances from the flooding. By the 1870s, however, it was clear that even these villages were not out of danger. When the Chesapeake & Ohio Railroad (C&O) sought to lay tracks from Memphis northward, George Millington and his wife, the former Miss Wade Bolton, subdivided a portion of their extensive landholdings on high ground for the construction of a depot and town. The depot was named Millington, and when villagers grew tired of fighting rising waters and moved closer to the new train station, the town of Millington was born.

The ability to transport lumber by train rather than over waterways made Millington businessmen wealthy even faster. Once lands were cleared, cotton became the crop of choice and the cotton industry employed untold numbers from the antebellum era through the 20th century. The railroad also brought new people to the town, and new businesses arose to accommodate their needs. By 1910, seven years after incorporation, Millington boasted a hotel, bank, sawmill, restaurant, livery stable, light plant, telegraph office, real estate agency, and newspaper. The town had nine dry goods and general merchandise stores, a drugstore, a jeweler, an electrician, three blacksmiths, six physicians, two undertakers (one female), four grocers (including two African Americans), and three barbers, one of whom was African American. In addition, Eugene Woods was a lumber manufacturer who employed dozens of men and created work for the town's three loggers and nine carpenters. Other residents found work with the railroad, in the slaughterhouse, as bookkeepers, as domestic or farm laborers, or in local schools. Entrepreneurial souls struck out on their own; Millington salesmen included peddlers of clothing, sewing machines, and harnesses.

In less than a decade, though, hostilities in Europe garnered much attention in the United States. To the east of Millington, one of West Tennessee's first pilot training facilities, complete with hospital, was developed at Park Field for the US Army Air Forces. Several landowners extended five-year leases to the government providing the land for Park Field, after which time the government purchased the land to make the airfield a permanent installation. One benefit of Park Field was that it placed Millington, and nearby Memphis as well, on an aerial mail route for the first time. Once the war ended, however, Park Field fell fallow until after the Depression, when it was temporarily revitalized as a transient camp for unemployed men under Roosevelt's Resettlement Administration. It was slated to be subdivided among small farmers who lost their properties during the Depression when World War II erupted, signaling major changes for Millington's future.

The military presence around Millington grew as the E.I. DuPont de Nemours Company was commissioned to build a munitions factory west of Millington, near the community of Cuba, to manufacture TNT and smokeless gunpowder to support Allied efforts during World War II. More than 5,000 acres were purchased by the government for this purpose, forever changing the physical landscape. The "powder plant," as it came to be known, altered the economic landscape as well. More than 14,000 men and women worked in the plant during World War II, forcing Shelby County to make drastic changes to the infrastructure. Roads were laid or improved to accommodate the increasing traffic to and from the plant. Utility services were installed or upgraded throughout the region. Trailer parks sprung up in and around Millington to combat the population explosion as workers migrated from hundreds of miles around in search of employment. Cotton lay unpicked in the fields below huge billboards for Memphis businesses, such as Bry's Department Store and Pantaze Drug Store, which boasted that they would cash powder plant paychecks for free. After the war, the powder plant was almost immediately closed and disassembled, leaving a scar on the landscape that remains unhealed.

By 1942, the federal government elected to expand Park Field for use as a Naval Reserve Air Base commissioned to train aviation cadets. Within a few short months, the reserve base was renamed the naval air station. Just east of the Millington city limits stood the world's largest inland Navy base, and for more than half a century, sailors and marines took instruction within its gates. The constant influx of sojourners was a great boon to the Millington economy, which, like the rest of the country, continued to recover from depression and war. While the base provided housing and recreation to enlisted men and officers alike, troops, no doubt eager for a few moments that felt like normal life, spent their paychecks at Millington's businesses and participated in Millington's community events. For many, the stay in Millington was brief—only a few weeks or months. Others stayed up to three years. With the added population of the Navy base, Millington's census numbers tripled by 1963, but more than half of those enumerated had no roots, historical knowledge, or significant attachment to the town they so briefly called home. In 1994, the first phase of the Navy Base Realignment and Closure (BRAC) Committee study resulted in a drastic change of assignment for the Millington base. After more than half a century, the naval air training facility became the Bureau of Personnel (BUPERS), causing a drastic decline in Millington's total population even as the area's per capita income increased with the influx of officers and reduction in enlisted personnel.

New Deal legislation and programs in the 1930s contributed to the development of the Shelby Forest Recreational Demonstration Area, west of where the powder plant would be built. Participants in Works Progress Administration (WPA) programs and local civic groups collaborated to carve out roads and woodland trails, plant 2.5 million trees, and build cabins and communal buildings in the new park. Three hundred WPA workers lived at Park Field while assigned to the Shelby Forest project. Residents of Millington and its outlying communities were proud to have this beautiful park so close to home and have enjoyed countless hours of hiking, swimming, fishing, camping, Frisbee golf, horseback riding, canoeing, and other activities there. Control of Shelby Forest, as it is locally known, was given to the State of Tennessee in 1944. It was renamed Meeman-Shelby Forest State Park in 1967 in honor of journalist and conservationist Edward J. Meeman.

Over the course of the 20th century, Millington welcomed progress. B.L. Howard purchased the fledgling Millington Telephone Company in 1928, and his grandchildren recently inherited control over this utility as well as the Millington Cable Company and the Old Timers Restaurant from their father, W.S. "Babe" Howard. The Strand Theatre, opened in 1939, offered second-run movies in its first life and now, in its second incarnation, welcomes crowds for live gospel and country music shows. By 1950, the Millington Post Office began door-to-door delivery of the daily mail. During the 1960s and 1970s, favorite pastimes included sipping fountain drinks at the Rexall or watching movies at the Twin Drive-In. The 1980s ushered in the era of Olympic baseball training in Millington at USA Stadium where, since 1990, another favorite event has been the International Goat Days Festival. Annual events such as the Flag City Freedom Celebration, the Mid-South Air Show featuring the Navy's Blue Angels, and the American Legion Fair draw large crowds. Although the closure of Memphis Motorsports Park in 2008 took NASCAR out of the Millington area, the 2011 opening of the Memphis International Raceway in its place offers enthusiasts the thrills of drag racing.

Today, Millington continues to thrive and prosper. The Millington, Sigler, Jones, Sneed, Armour, Howard, and other early families continue to branch out of the town's deep roots. Former military personnel have returned and raised families in Millington. Many who have grown up in and around Millington have gone on to fame and fortune and have given back graciously to their hometown community. Pop star and actor Justin Timberlake grew up near Shelby Forest, where he has recently opened the world-class Mirimichi Golf Course. Singer-songwriters Bobby Whitlock, Myla Smith, and members of the band Kings of Leon grew up in and around Millington. The Reverend Al Green, a gospel singer and pastor at Memphis's Full Gospel Tabernacle Church, also calls Millington home. Football players Tyrone Calico (Tennessee Titans), Greg Hardy (Carolina Panthers), Ahmaad Galloway (Denver Broncos), and Marlon Barnes (Colorado Buffalos) started their careers as Millington High School Trojans, and former University of Memphis Tigers defensive back Chris Michaels has returned to his alma mater as head coach of the Millington Trojans. During USA Stadium's heyday, over 200 major league baseball players trained for Olympic competition over the course of a decade. Millington, as its founding families knew more than a century ago, their descendants continue to appreciate, and newcomers are beginning to realize, is a great place to raise a family. The photographs in the following pages reflect the way Millington once was and how it has evolved into a fine place, as welcome signs at the city limits invite, to "Come Home" to.

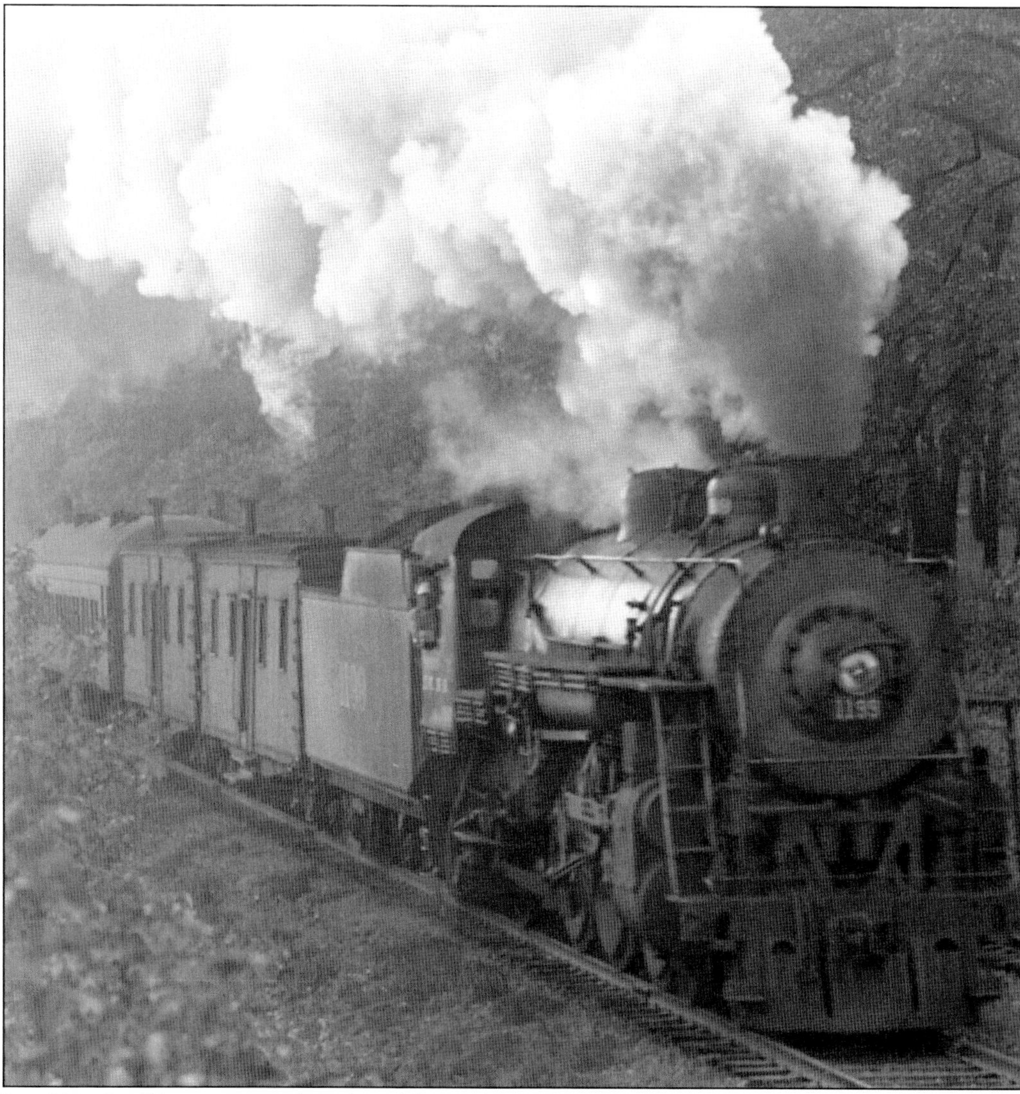

In 1896, the Illinois Central Railroad (ICRR) purchased the Chesapeake & Ohio line (C&O), which passed through Millington. After World War II, a coal strike prompted the ICRR to replace its steam engines with diesel engines. By the late 1950s, passenger service had sharply declined and even mail service was in danger of extinction along the ICRR lines. The steam engine known as "Old Hoodlum," pictured here, made its final run on December 5, 1947.

One

THE MILLINGTON COMMUNITY

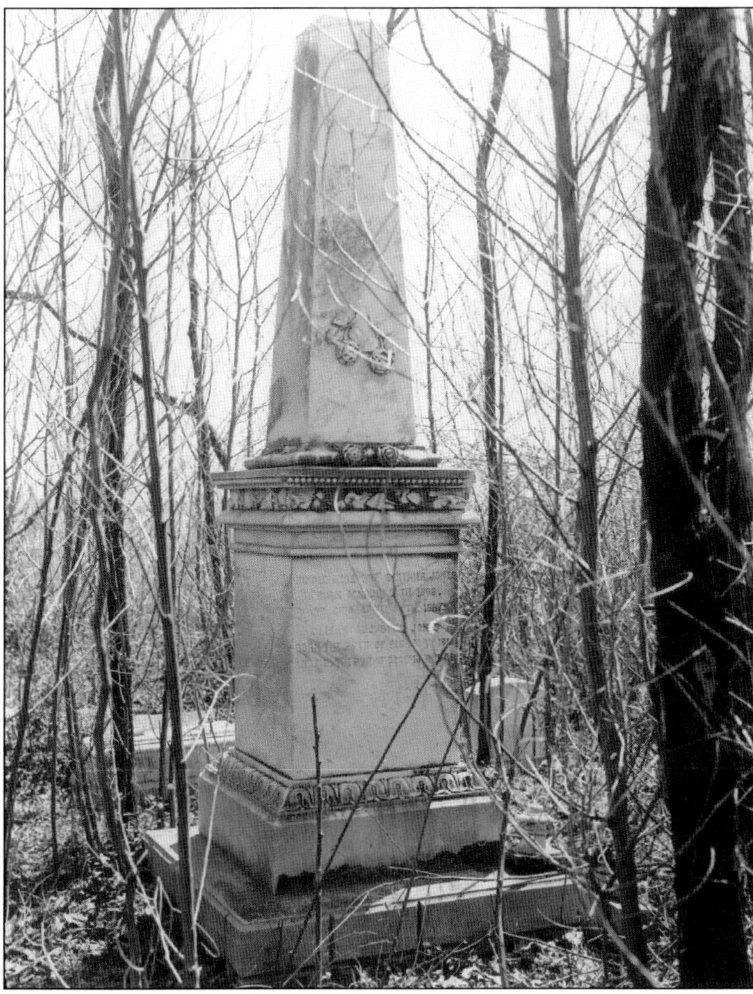

In 1829, the heirs of Gen. Chamberlayne Jones of Amelia County, Virginia, established one of Shelby County's wealthiest plantations along Big Creek. The Shipp (or Jones) Cemetery monument states that they were "the first settlers of this place." The cemetery is at the end of a runway on former naval base property, prompting a Navy photographer in 1959 to label this image "Graveyard Aboard." (Official US Navy photograph; courtesy UM.)

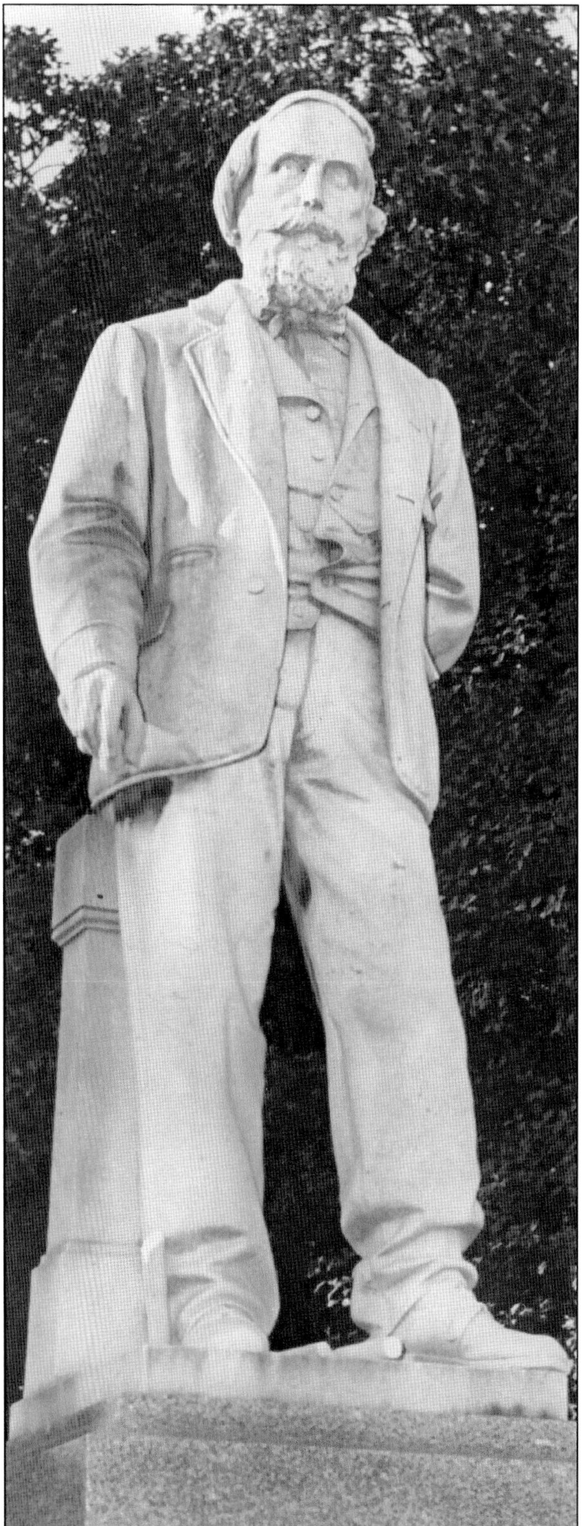

An infamous slave trader and wealthy eccentric, Wade H. Bolton left a fortune to his niece and namesake, Mrs. Wade (Bolton) Millington, wife of Dr. George H. Millington. Before moving to Texas, the Millingtons offered part of their Glencoe plantation for construction of a train station to be named for them and subdivided the property surrounding the depot. It is interesting to note that George Millington's father, Thomas H. Millington, was the chief engineer for the Mississippi River Railroad and was tasked with finding a straight and direct line from Memphis to Covington, about 25 miles north of Millington. Neighbors from Millwood and Glencoe moved closer to the new Millington depot, which was built on high ground safe from the flooding that plagued the area. The Town of Millington incorporated in 1903.

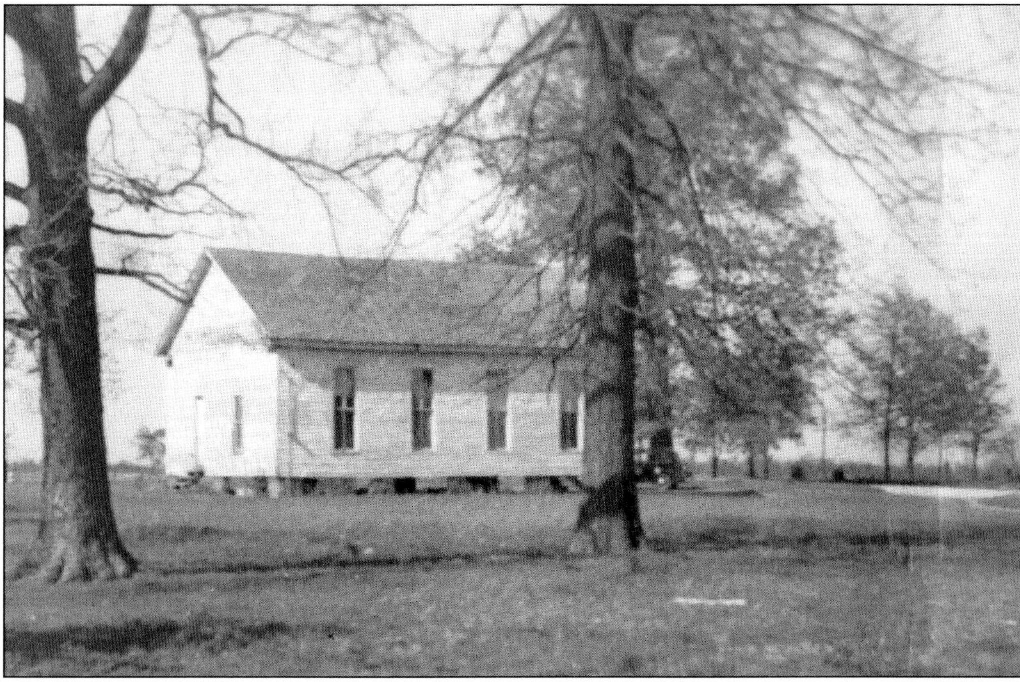

Big Creek Baptist Church is thought to be the oldest church in Shelby County. Located near Lucy, the congregation first met as early as 1812 in a brush arbor. A fire destroyed the first log building, and a box church was constructed in 1832 with sides added for slaves who attended services there. The church's third building, a frame structure shown above in 1937, was constructed in 1866. Below is the fourth church building, a concrete block structure constructed in 1944. In 1986, the church added a new sanctuary. Foreign missions have been an important part of the church's mission, furthered by the numerous military personnel who have attended over the years.

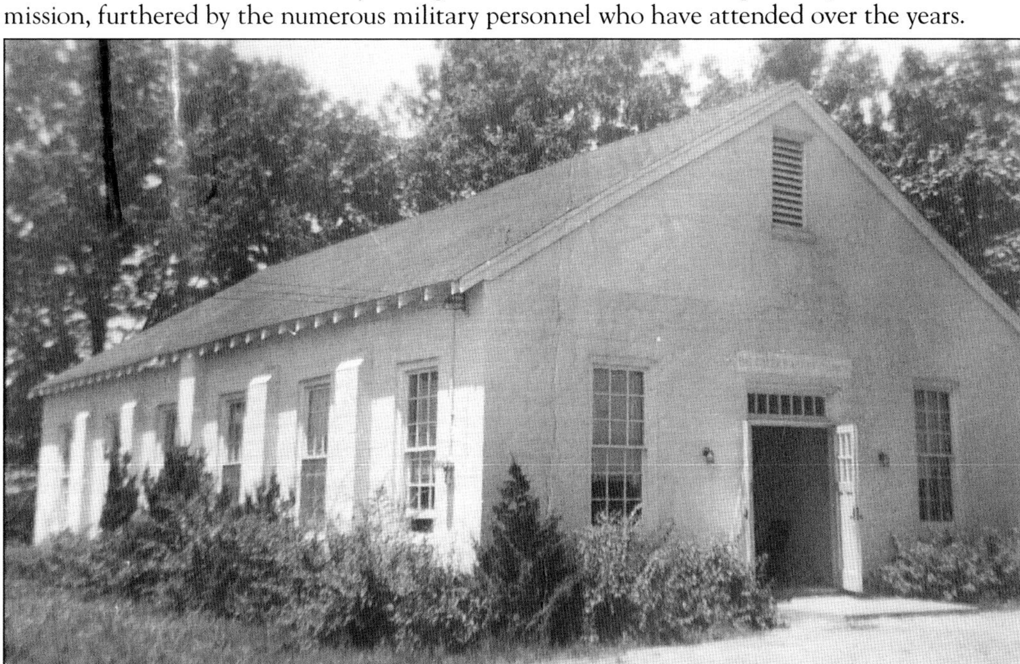

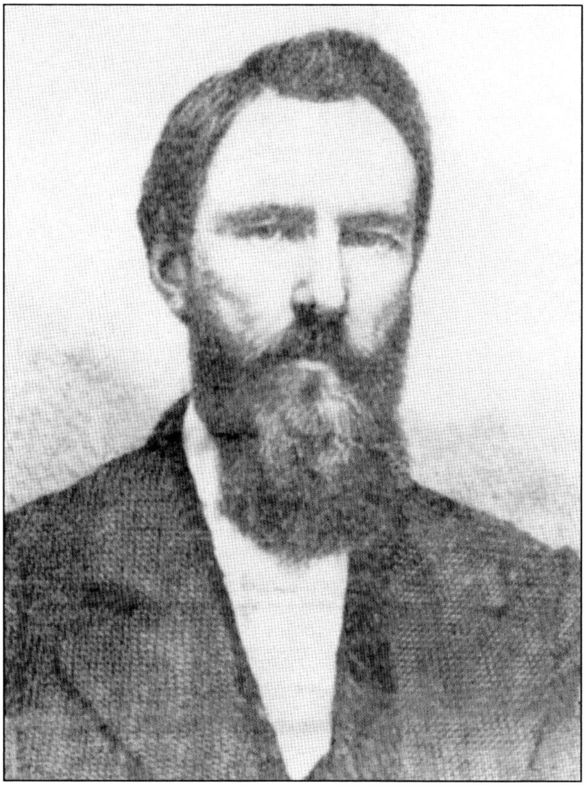

Samuel S. Rembert III was the grandson of one of the earliest settlers of Woodstock, once known as Rembert Town, where one Rembert home built in 1842 is still occupied by a descendant. Rembert spoke and wrote on matters of spiritualism and invented a number of mechanical innovations, including a wire buggy wheel, a cotton press, and, long before the advent of the refrigerator, a method of ice making. In 1850, Rembert and Jedediah Prescott patented the first mechanized cotton picker. Rembert's son convinced them not to produce the contraption for fear of idleness among slaves. When the image below appeared in the *Memphis Press-Scimitar* in 1937, the debate continued over detrimental effects on farm laborers caused by International Harvester's modern cotton equipment. By the 1970s, manual cotton picking had been largely replaced by mechanical methods.

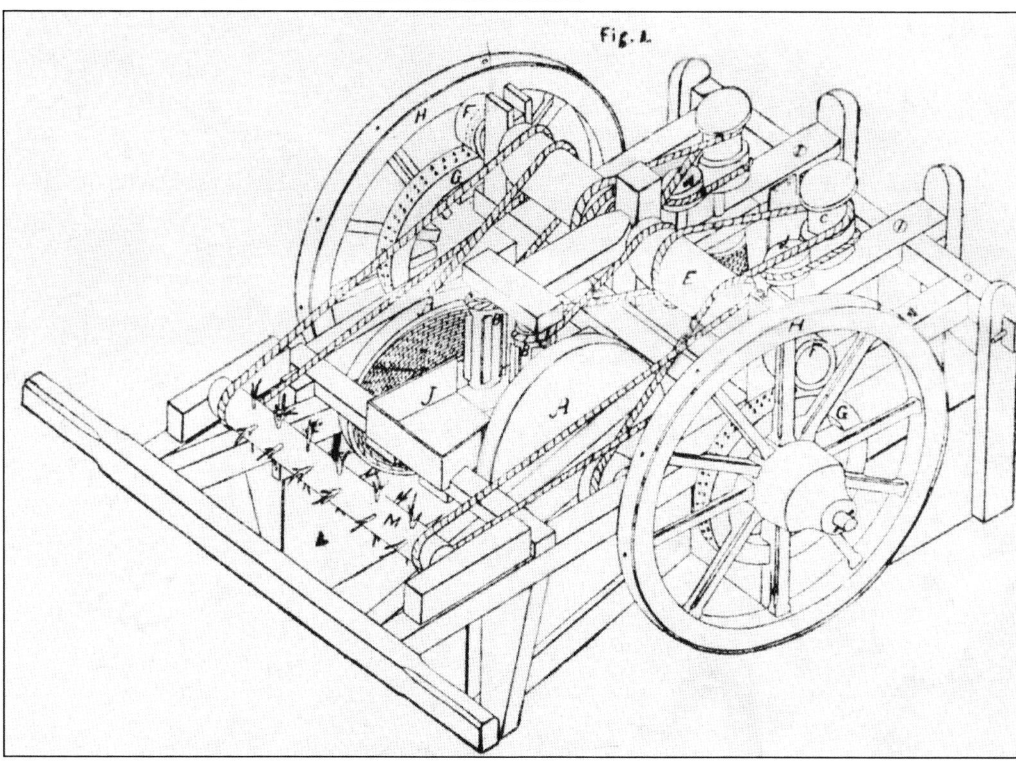

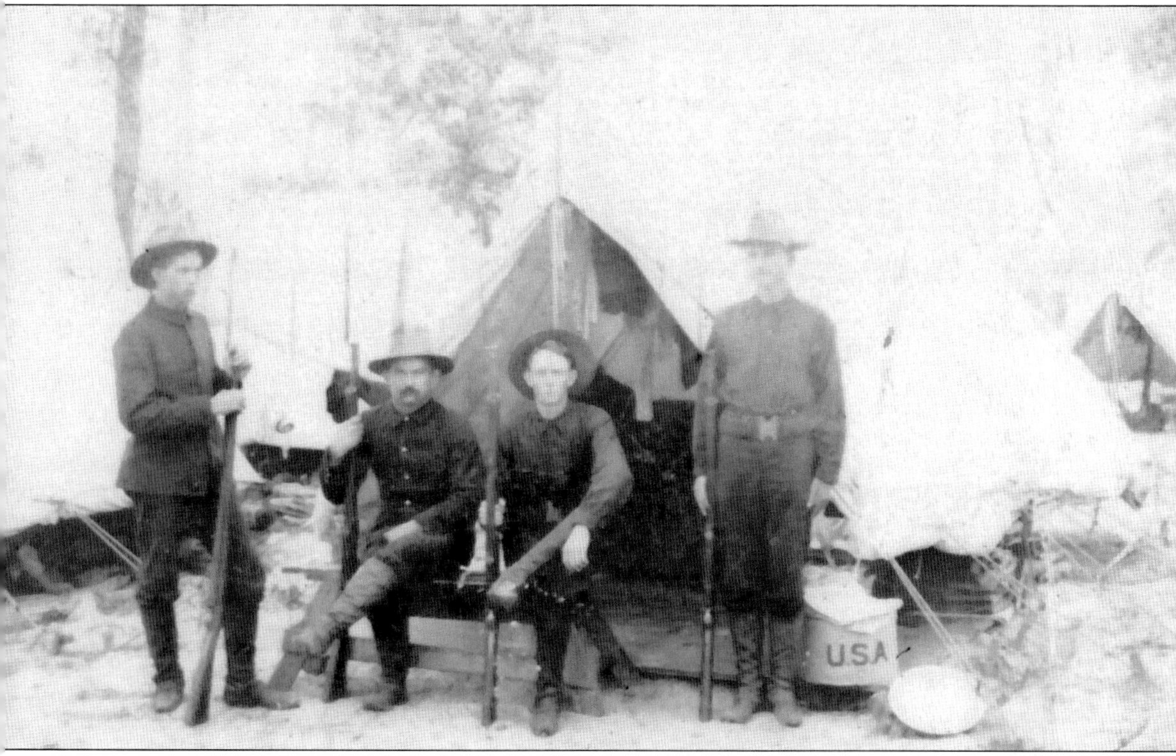

William Lee Monasco (second from left) served in the Spanish-American War. He enclosed this photograph in a letter home, dated July 19, 1898, telling his family, "I am going to Cuba without doubt." His sister Jennie Harmon lived in Millington. He was discharged in December 1898 after an injury to his hip. A Tennessee native, he settled on a farm in McAlester, Oklahoma. Monasco Road in the bluff area is named for his family. (Courtesy MTC.)

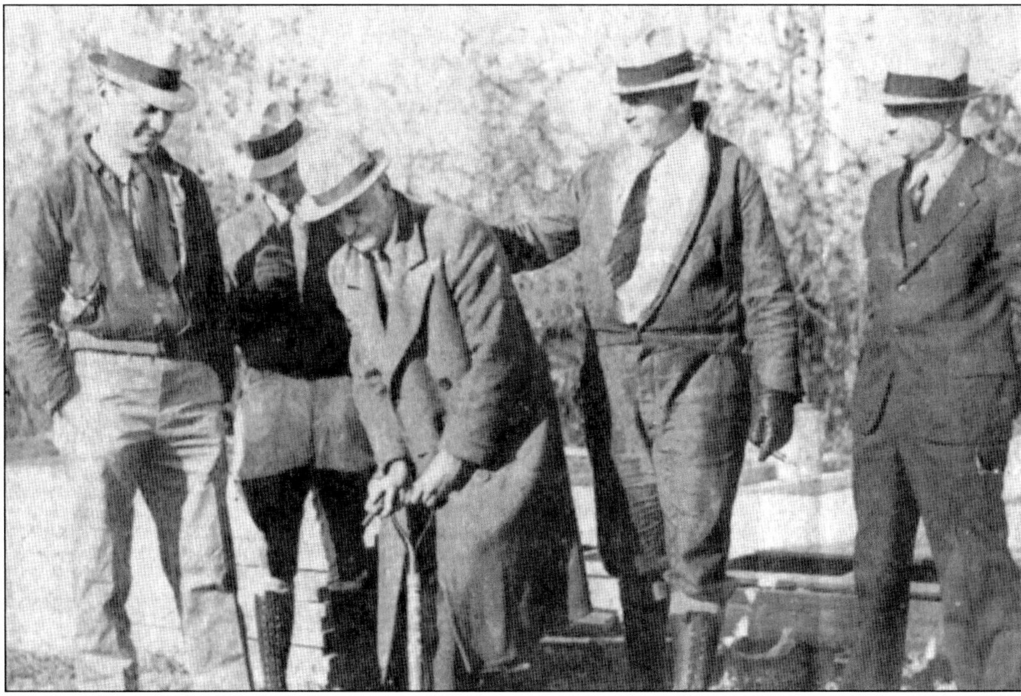

Squire Emmet Early Jeter (above, center) was a justice of the peace, county court chairman, magistrate, Tennessee congressman, and respected businessman. He was instrumental in procuring land for the development of Shelby Forest, where he turned the first shovelful of dirt for Jeter Lake in 1935, as pictured above. With him are, from left to right, engineer H.W. Smith, the squire's son and junior foreman Paul Jeter, foreman E.L. Beckham, and Smith's assistant C.B. Parsons. Before construction of the park, the Jeter community included the E.E. Jeter School and at least two stores, including Peter's Shoes, shown below. (Above, courtesy James Clifton and Jamie Scholik; below, courtesy Doug Ammons, Shelby Forest General Store.)

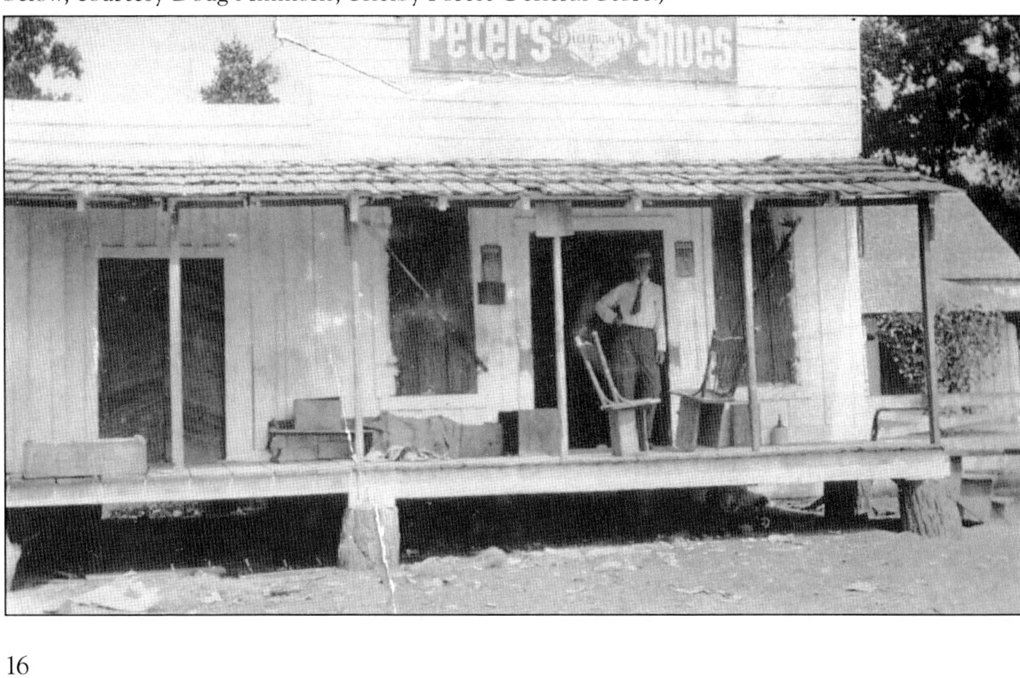

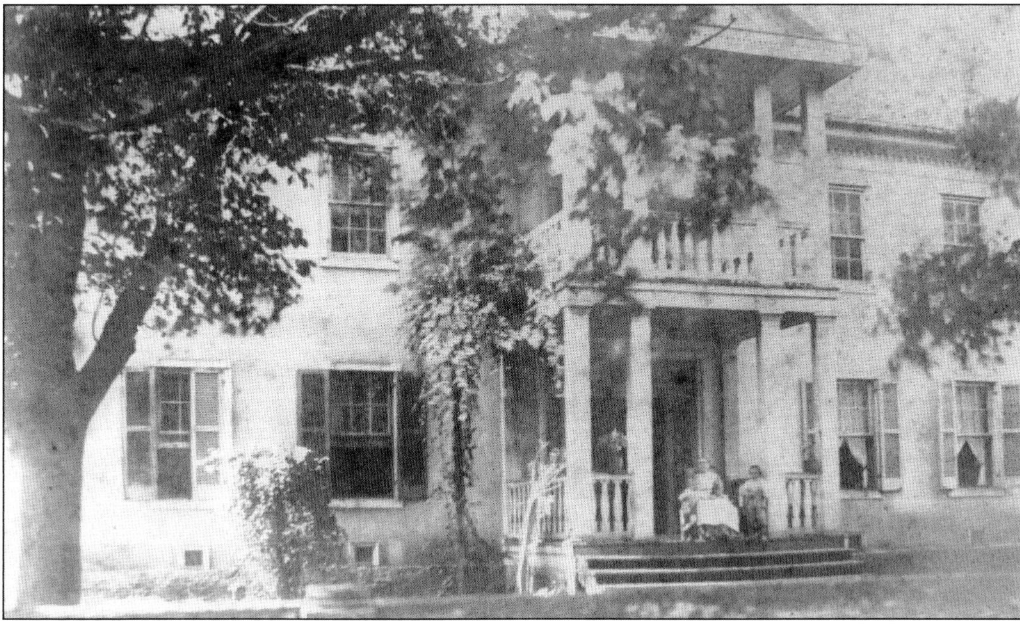

Antebellum architecture, seen above, often had elements of Georgian, Federal, Colonial, and Greek Revival styles. Plantation homes around the Millington area, such as the 1842 Rembert home near Woodstock, the Williams home near Lucy, and the Stovall home near Cuba, for example, were typically two-story homes with columned verandas, front balconies, central entryways with sidelights and transoms, and symmetrical double-hung windows flanking the entry. Of these, only the Rembert home is still standing and remains in the possession of descendants. Early homes in Millington proper, like the one below, were built later, usually in the Victorian style. There are very few examples of Victorian architecture remaining in Millington. (Above, courtesy James Clifton and Jamie Scholik; below, courtesy MTC.)

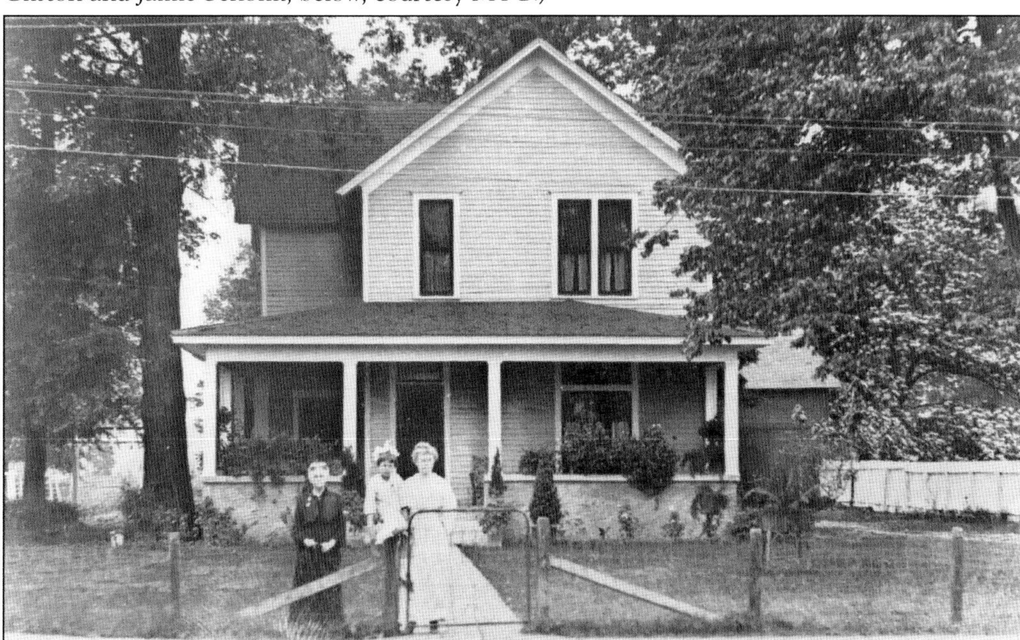

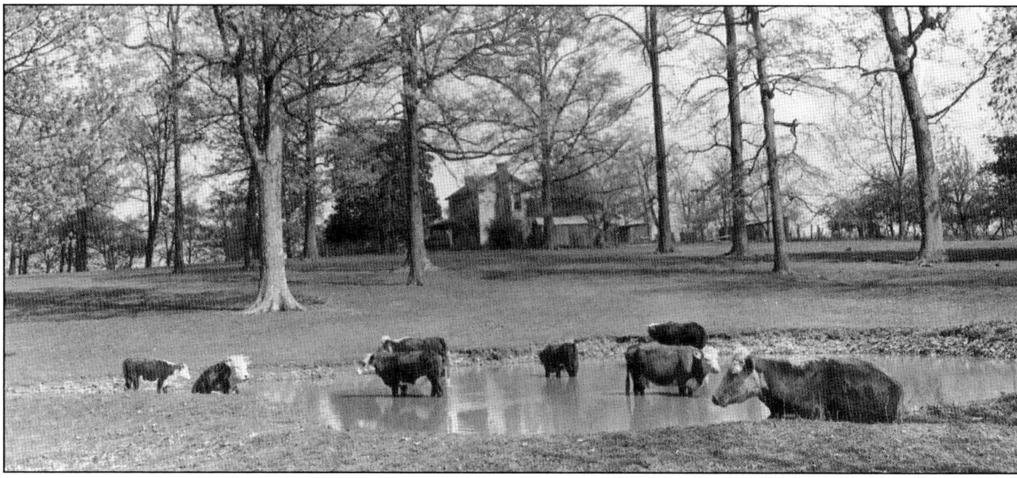

Dr. James Sedberry of Tipton County, Tennessee, bought property in Shelby County in 1867. There he built this home, barely visible through the trees, and raised his family. His daughter Jennie married John Aycock, a representative of the Shelby County Board of Education from 1912 to 1934. She and her husband raised a family in her childhood home, where she lived for over 75 years before the Navy took over the property in the 1940s.

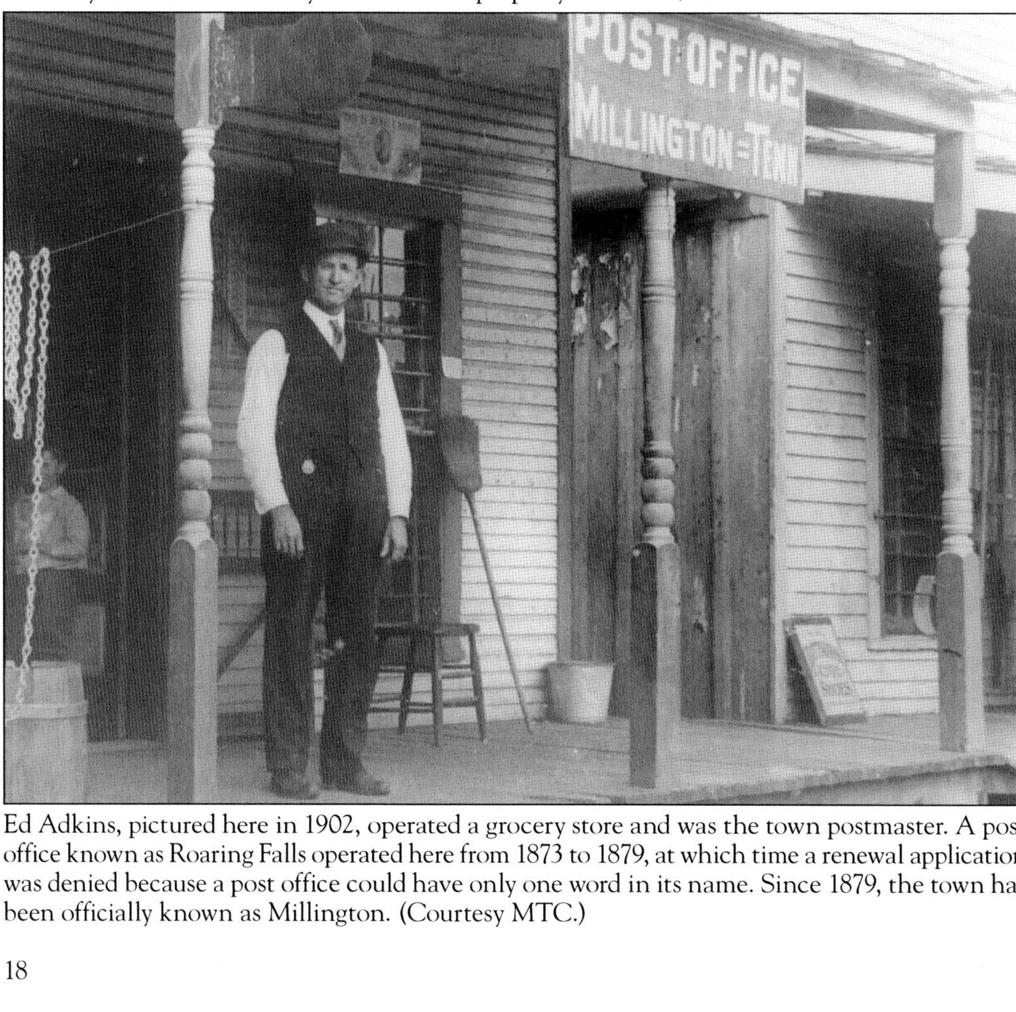

Ed Adkins, pictured here in 1902, operated a grocery store and was the town postmaster. A post office known as Roaring Falls operated here from 1873 to 1879, at which time a renewal application was denied because a post office could have only one word in its name. Since 1879, the town has been officially known as Millington. (Courtesy MTC.)

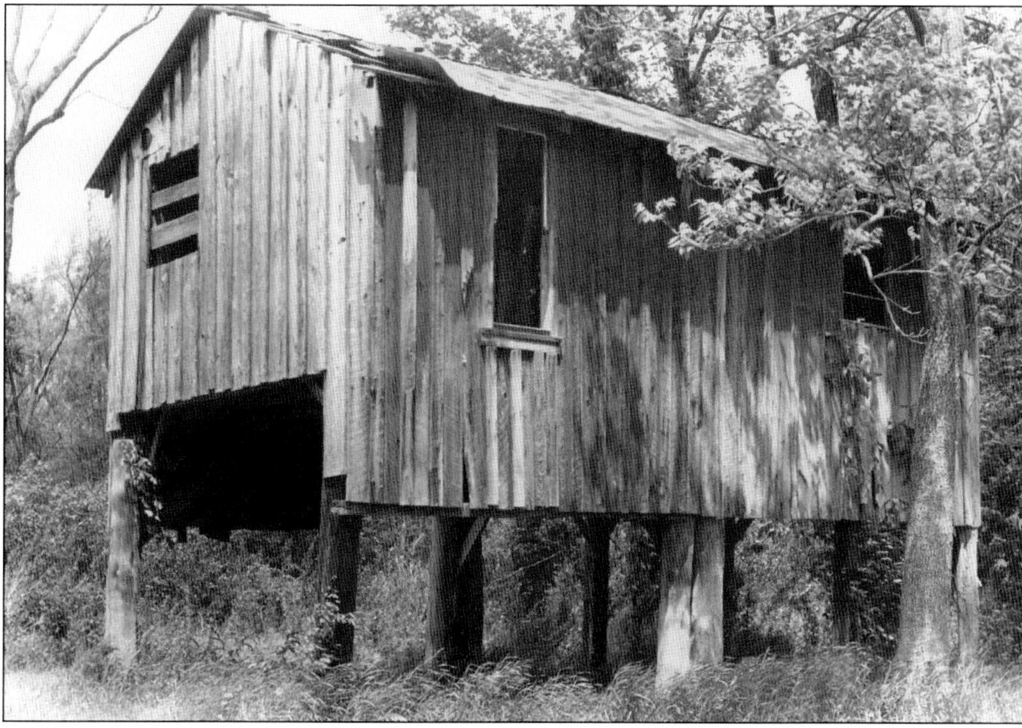

Of course, not all residential buildings were as grand as the plantation and town homes of wealthier citizens. These cabins stood in the Shelby Forest area until at least the 1980s. Stilts, or piers, elevated these and other bluff area structures above the annual threat of rising floodwaters from the Mississippi and Loosahatchie Rivers and their tributaries.

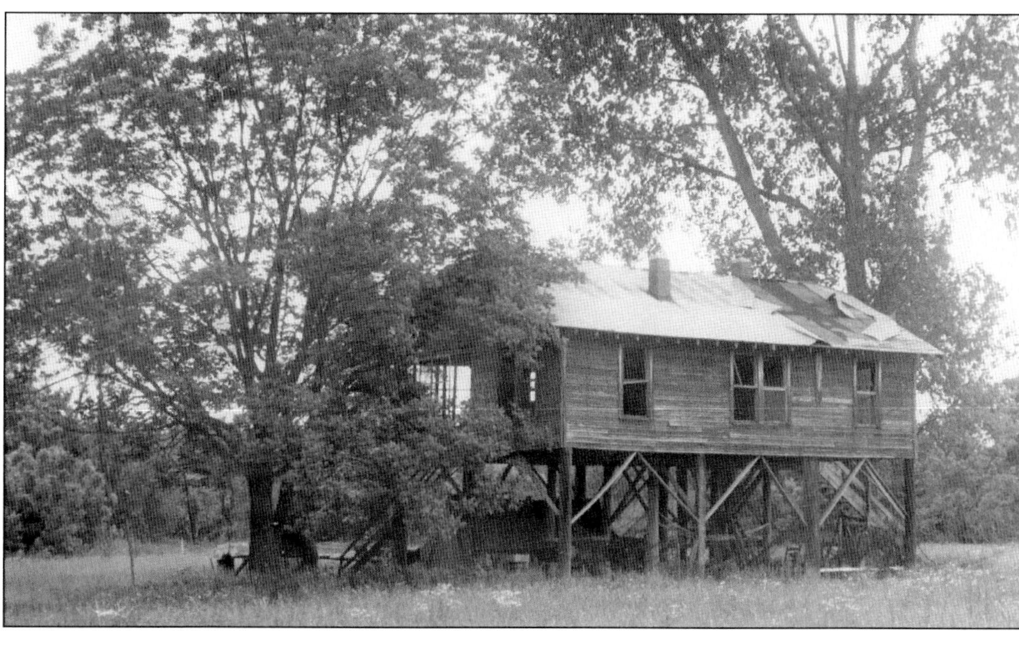

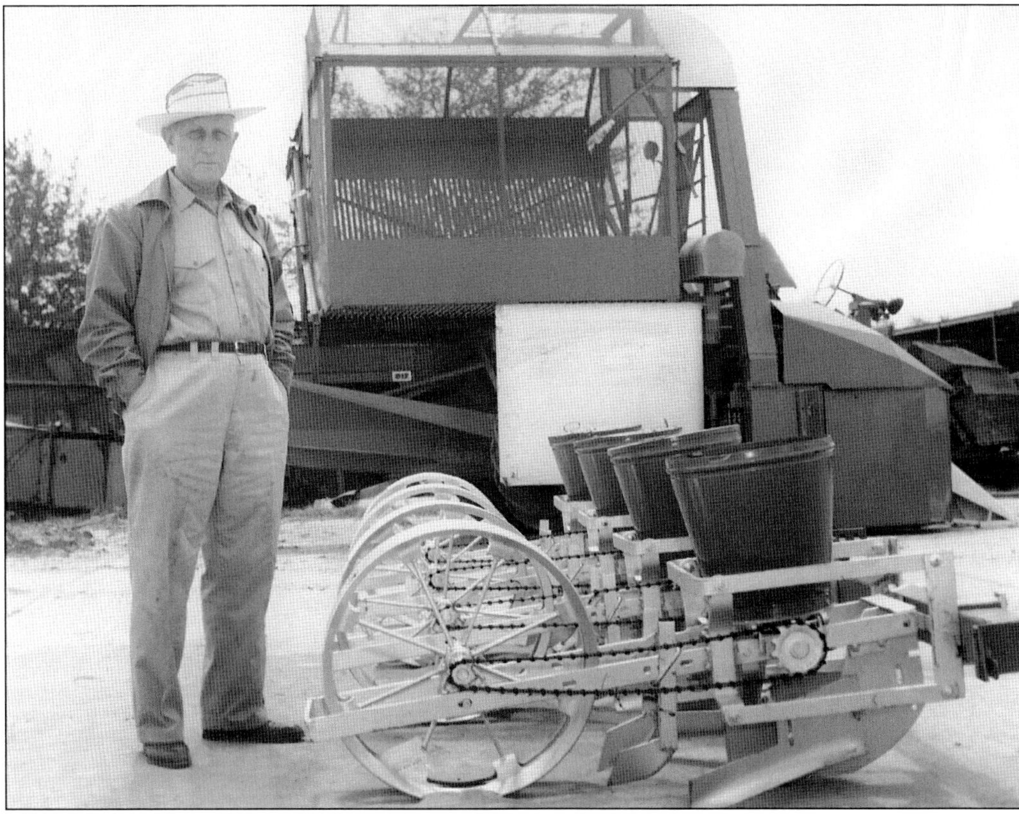

Claire Bert "Doc" Lucado, shown above in 1963, was one of the landholders whose properties later comprised the Navy base. Lucado also maintained acreage west of Millington and in Tipton County, including Island 36 or Lucado Farms, which had belonged to his grandfather, William Sigler, since the Civil War. Below, due to its location on the Mississippi River, Lucado Farms was flooded frequently enough that Lucado built a dwelling on stilts. In dry weather, he operated a store out of the lower level, which he enclosed with cinder blocks in later years. The dwelling house is the small two-story building in the center of the photograph.

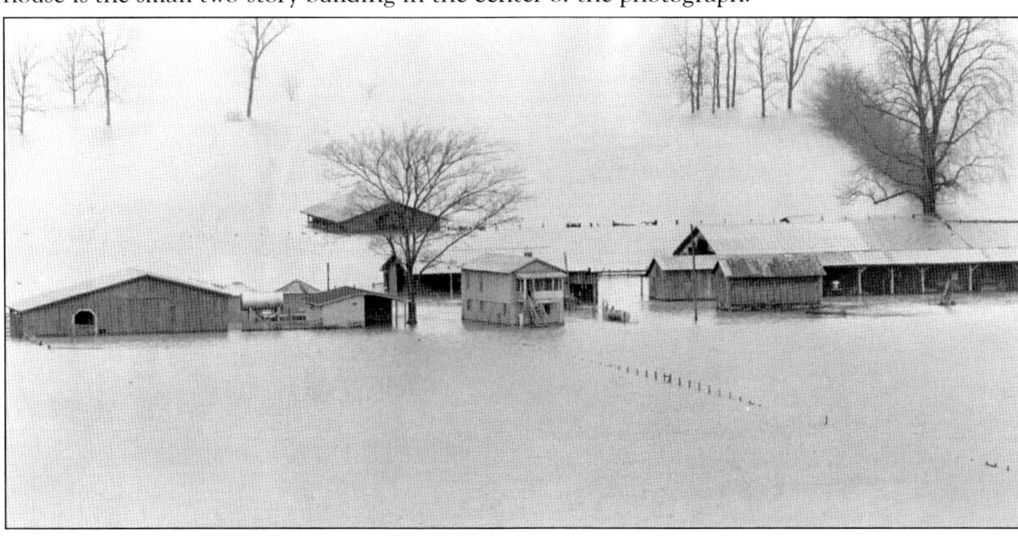

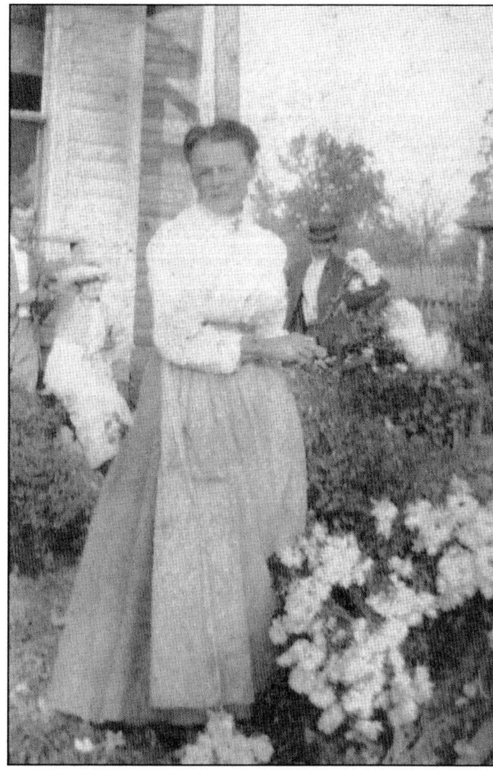

Mary Ellen "Mauzie" (Sigler Lucado) Scott was the oldest child of William and Martha Carrel Sigler. She and her sister Alva donated land in 1880 for the Ashland School. Charl Williams, in the white dress below, had her first teaching job at Ashland, which closed in 1907. Williams was superintendent of Shelby County Schools from 1915 to 1922 and later became the National Education Association president. Many one-room Shelby County schools were reorganized, consolidated with other schools, or closed altogether during the early 20th century as Williams and the school board strove to improve standards for education in rural areas. It was during this time that a central high school was established for each district. Children from Ashland transferred to Cuba School and then to Millington Central High School. (Right, courtesy James Clifton and Jamie Scholik; below, courtesy SCA.)

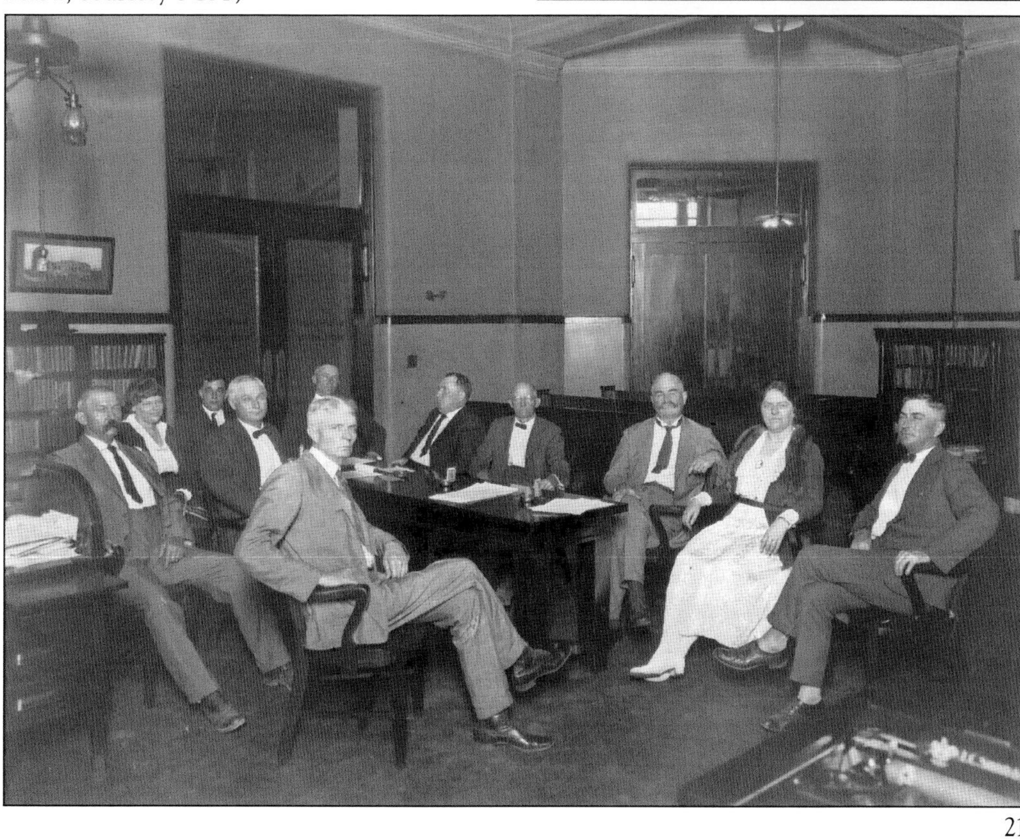

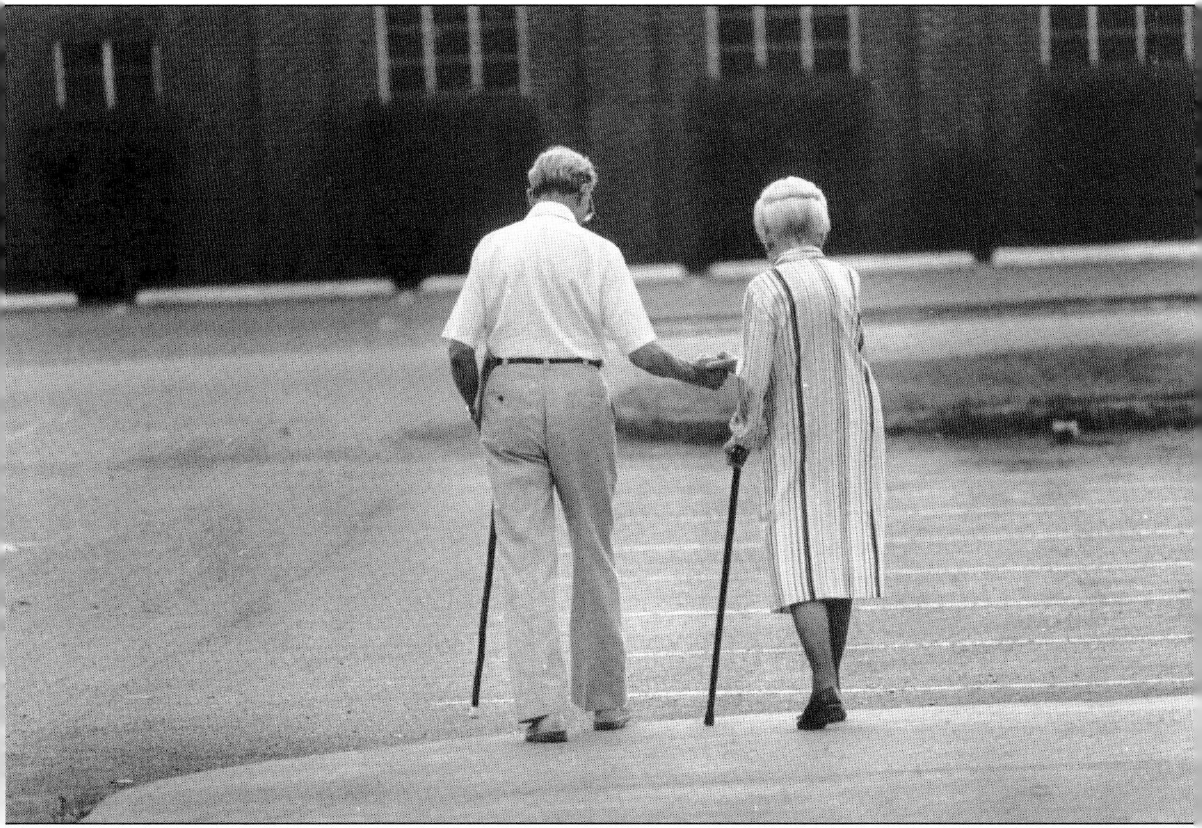

In 1927, James Williams married Thelma Sigler, merging two of the area's oldest families. The only child of Millington merchant Buck Sigler, Thelma was a teacher at Millington Central High School. She wrote *The Early History of Millington* and a history of First United Methodist Church. Her husband was also the son of a Millington merchant, James Williams Sr., whose store was a forerunner of the E.A. Harrold store. Her mother-in-law, Daisy Williams, was E.A. Harrold's sister. James Williams Jr. was a civil service employee at the naval air station for 42 years. By the 1980s, they were one of Millington's oldest native couples. They donated land at the corner of Highway 51 and Wilkinsville Road for Kiwanis Park in 1981. Thelma passed away in 1991 after 64 years of marriage. That year, James was honored on his 94th birthday as Millington's oldest native resident. He lived another four years.

Frank C. Ryburn was a Navy pilot and instructor during World War II and once owned a Ford dealership in Millington, but it is his work with the Memphis and Shelby County Industrial Board and the Millington Industrial Development Board for which he is best known. Ryburn died in 2008 at the age of 85.

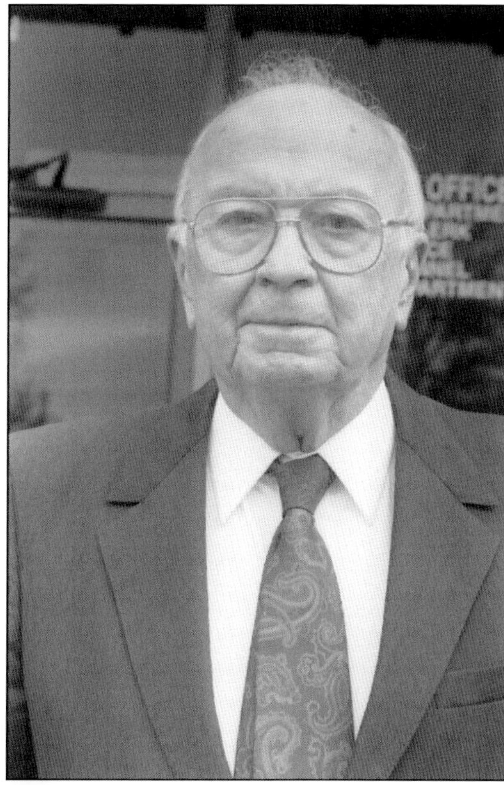

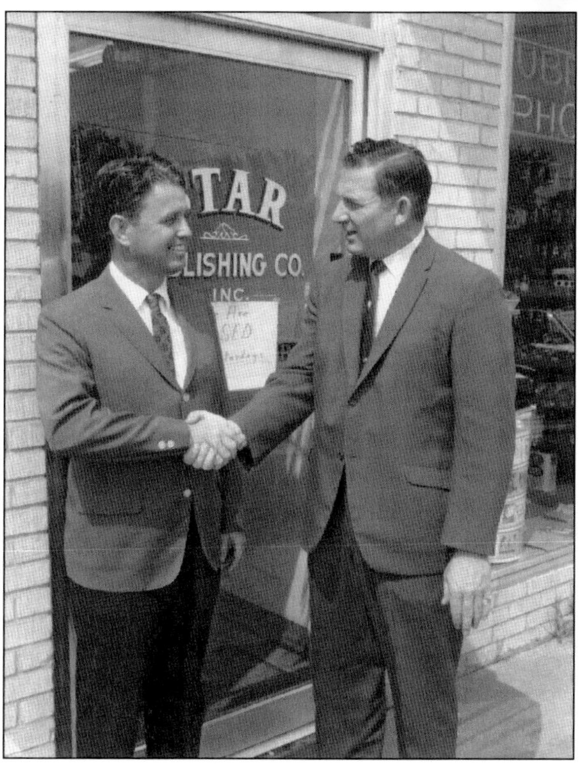

Pictured here in 1966, Leonard C. Dunavant (left) was a 1937 graduate of Millington Central High School and became a Millington alderman. He became a Republican state senator known for his principled politics, particularly on hot-button issues like prostitution, Medicaid reform, gambling, and the institution of a state income tax. Sticking to his guns during a 1989 campaign he declared, "I'd rather be responsible than re-elected!" (Courtesy *Millington Star.*)

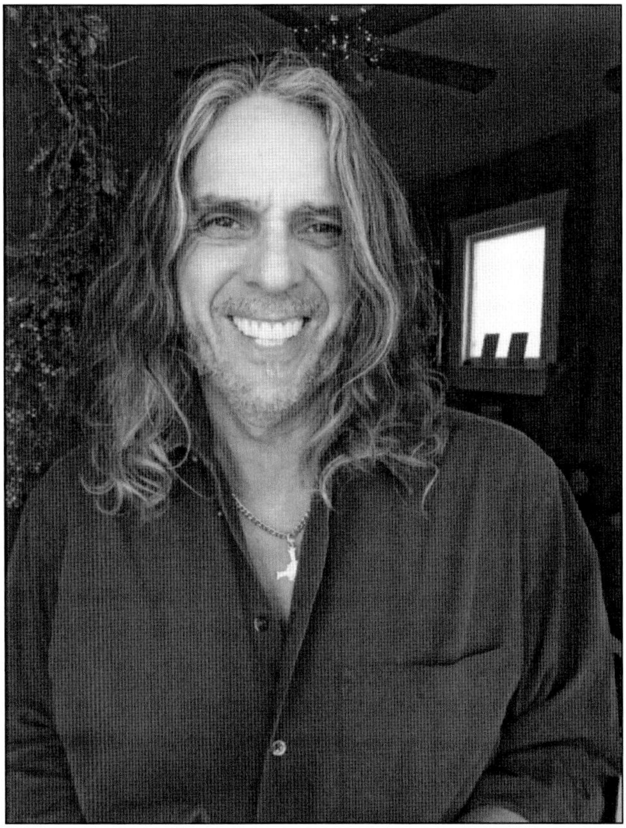

Singer-songwriter Bobby Whitlock was signed by the Stax Hip label at age 16. He has worked with Steve Cropper, Donald "Duck" Dunn, Delaney and Bonnie, Eric Clapton, George Harrison, Derek and the Dominos, George Jones, and others. Whitlock's recent autobiography recalls his childhood in Millington, including the day he unwittingly entered and won a childhood beauty pageant by joining a procession of children circling the Oak Park wading pool. Lee Waits Miles, pictured below in front of the wading pool, was the overseer at Oak Park during the summer months and the physical education instructor at Millington Central High School during the academic year. It was estimated that over 300 children visited Oak Park daily during the summer of 1961. (Bobby Whitlock image courtesy CoCo Carmel.)

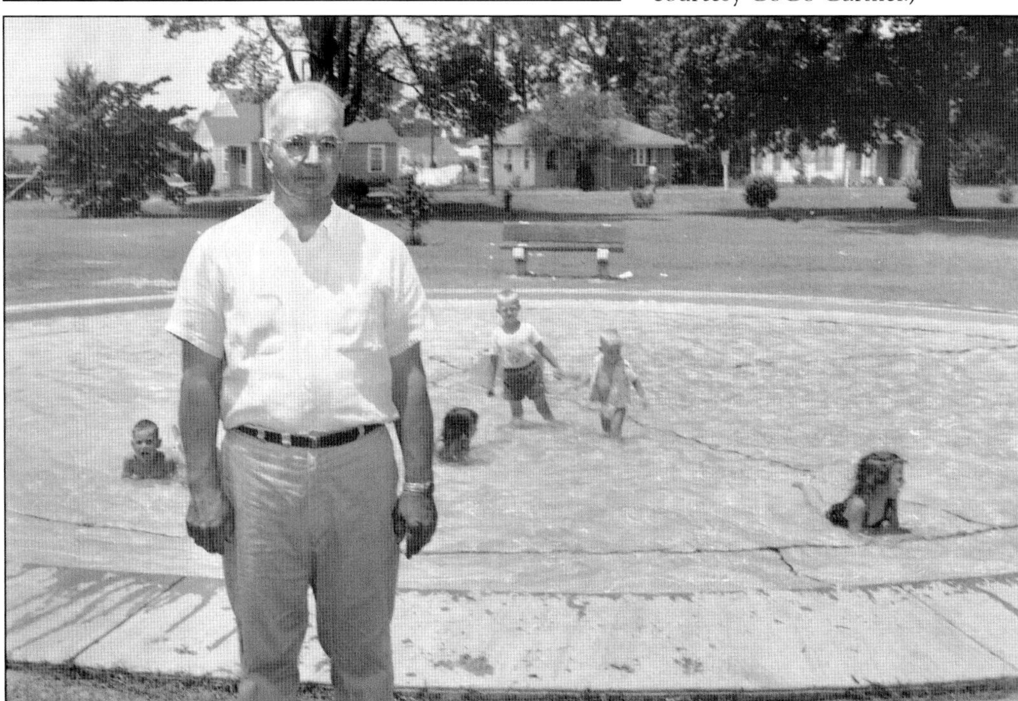

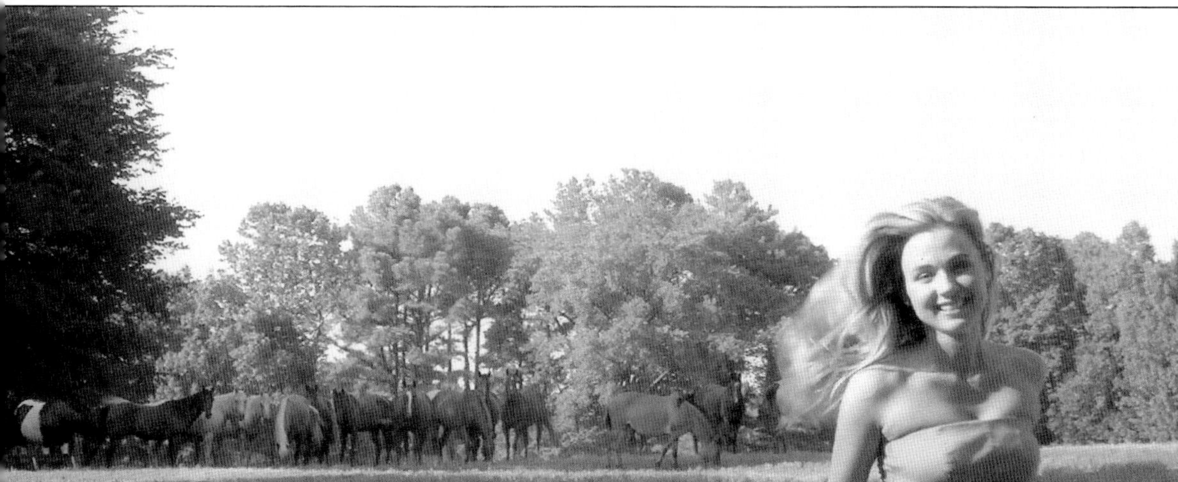

Singer-songwriter Myla Smith's fancy dress and bare feet hint at the old saying that "you can take the girl out of the country, but you can't take the country out of the girl." Smith spent her early years on a farm at Shakerag, a few miles west of Millington, where this photograph was taken. As a teenager, she appeared on the children's television program *Barney and Friends*. After recording some other projects, Smith was a high achiever at the University of Memphis. She has earned several awards since her debut album in 2006 and has since launched her independent record label, Shake Rag Records. Pop icon Justin Timberlake, another Millington native, also started his career in children's television on *The Mickey Mouse Club* before launching his musical career with N'Sync and finally going solo. Timberlake also stays true to his roots, as reflected by his William Rast clothing line, which got its name from Timberlake's grandfather William Bomar and cofounder Trace Ayala's grandfather, John Rast. Bomar and Rast are long-standing families in North Shelby County. (Courtesy Myla Smith Thomas.)

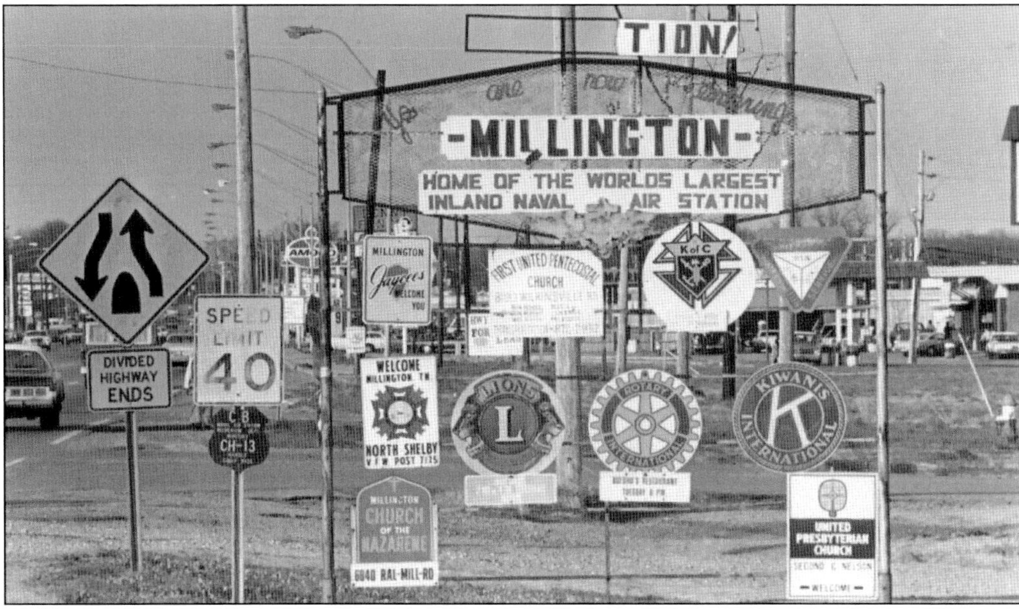

Visitors to Millington once were greeted by this sign that made it clear that civic participation was a fundamental part of the town's character. The United Presbyterian, First United Pentecostal, and Nazarene Churches were relative newcomers to the area's religious denominations. Civic organizations included the Lions Club, Kiwanis International, the Knights of Columbus, the Jaycees, the Rotary Club, and the Veterans of Foreign Wars. Buford's Restaurant hosted many of the meetings of these clubs. The sign, which had seen better days by the time this photograph was taken in 1980, no longer greets visitors, but the organizations still do. In 1972, an unexpected group of citizens used signs to take a stand on participation in Vietnam, the United Nations, and the Constitution in support of Col. Arch Roberts, an outspoken critic of the United Nations. (Below, courtesy *Millington Star.*)

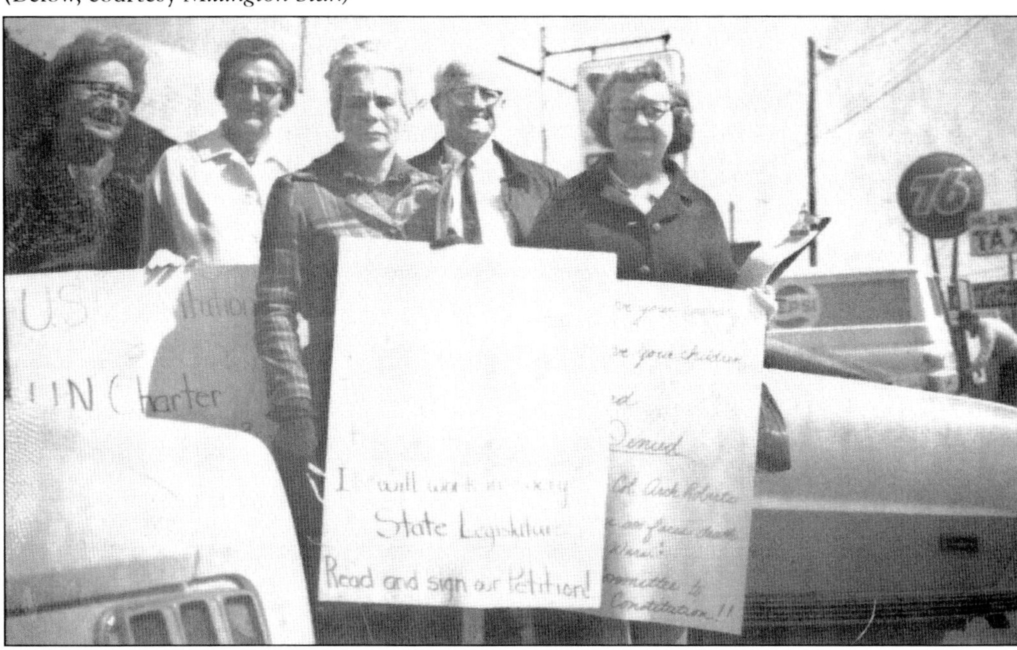

Other citizens were more private but no less devoted to making their voices heard. Here, Jessie Kinton Sigler votes for her choices for local and county government offices. (Courtesy *Millington Star.)*

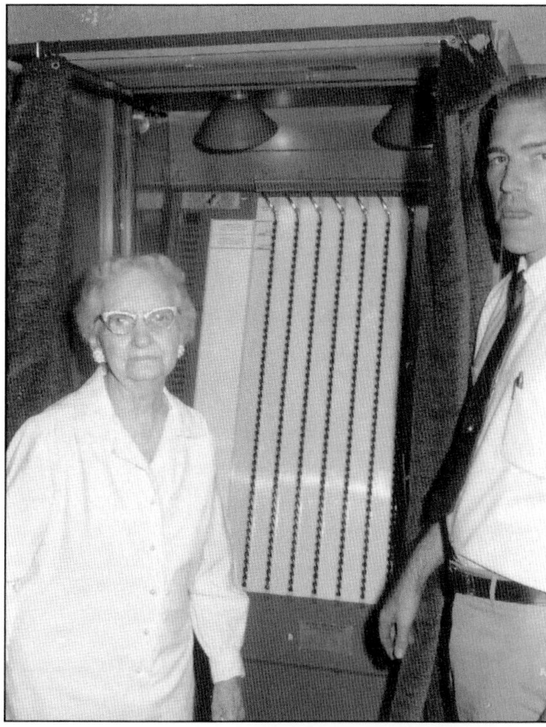

Below, members elected to the 1959 Shelby County Court from the Millington area included Paul Barret (third from left) and Charles Baker (fourth from left). The Barret family had a long history in the county courts and owned a store, cotton gin, and the Barretville Bank. Highway 385 is named Paul W. Barret Parkway in honor of his contributions. Charles Baker was chairman of the court in 1962 and a mayor of Millington. Charles Baker Airport and Baker Civic Center were named for him. (Courtesy SCA.)

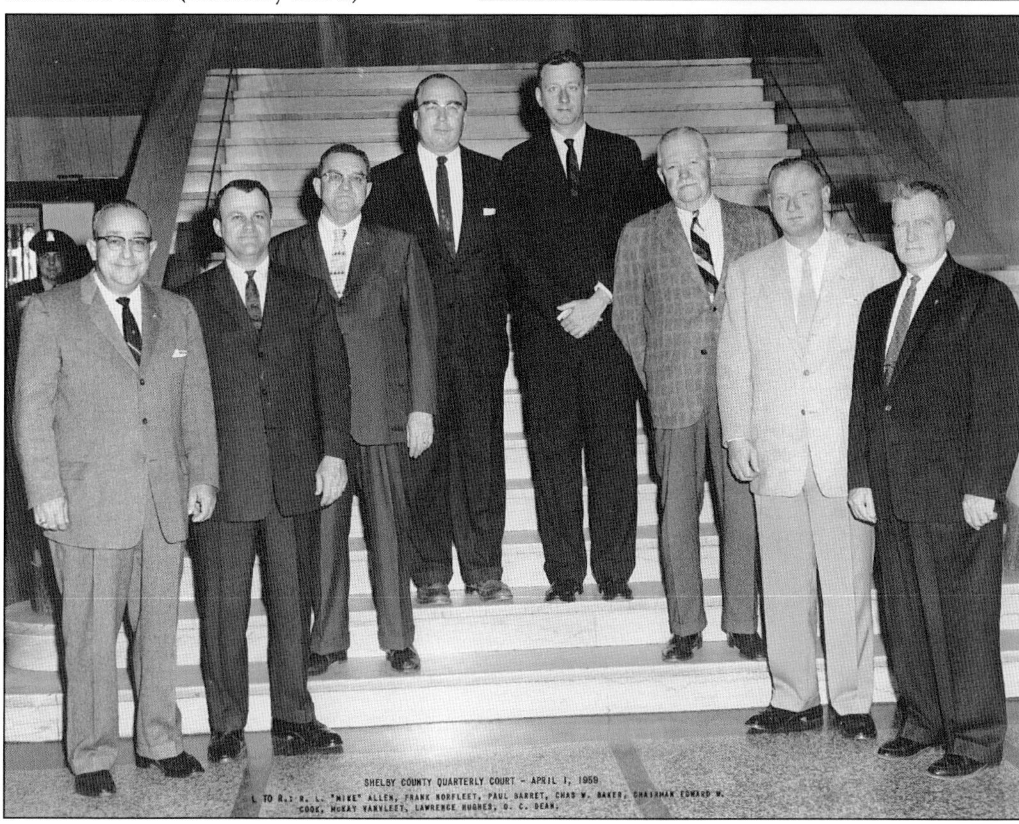

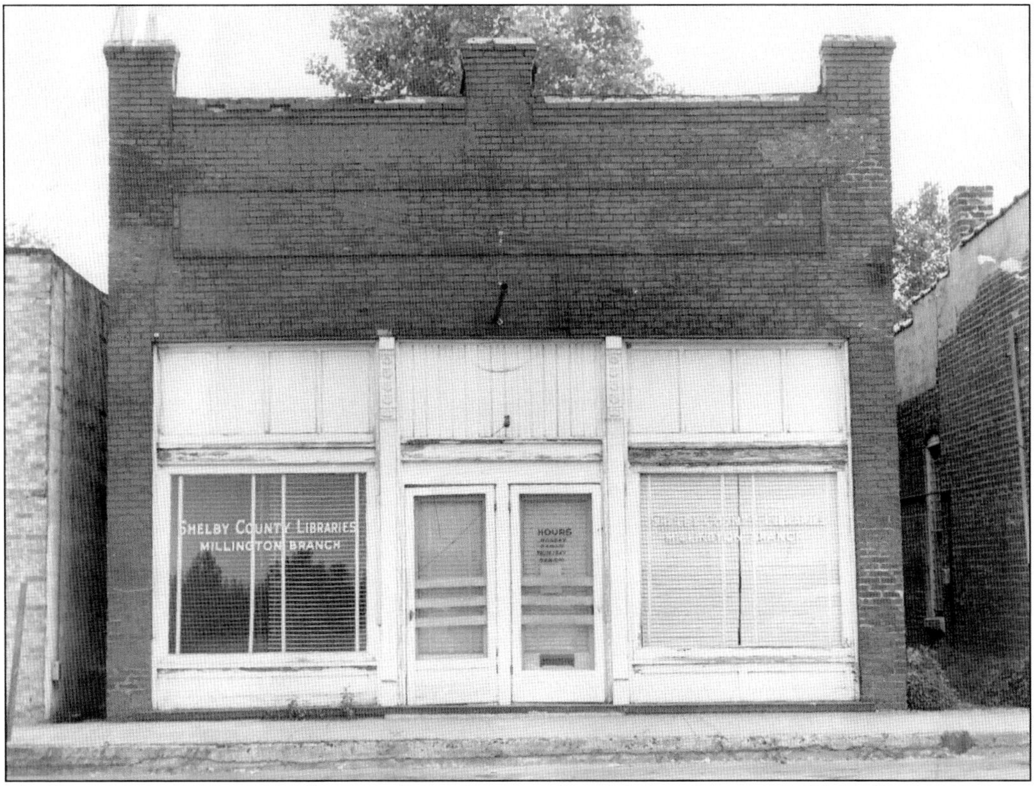

Shelby County Library's Millington Branch opened on Easley Street in 1956. The library was open only Mondays and Thursdays, and a slot in the bottom of one door served as a book drop when the library was closed. An early home to Millington City Hall, the Veterans of Foreign Wars purchased the building in 1961 to expand its lodge, and the library temporarily operated out of Baker Civic Center. A new city hall building was constructed in 1954.

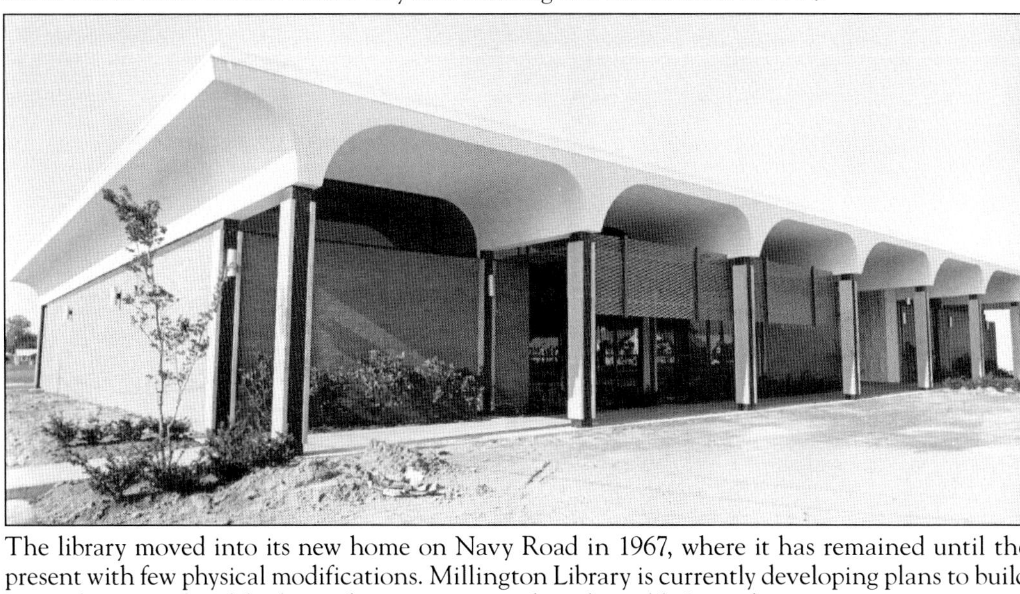

The library moved into its new home on Navy Road in 1967, where it has remained until the present with few physical modifications. Millington Library is currently developing plans to build a new, larger, updated facility to better accommodate the public's needs.

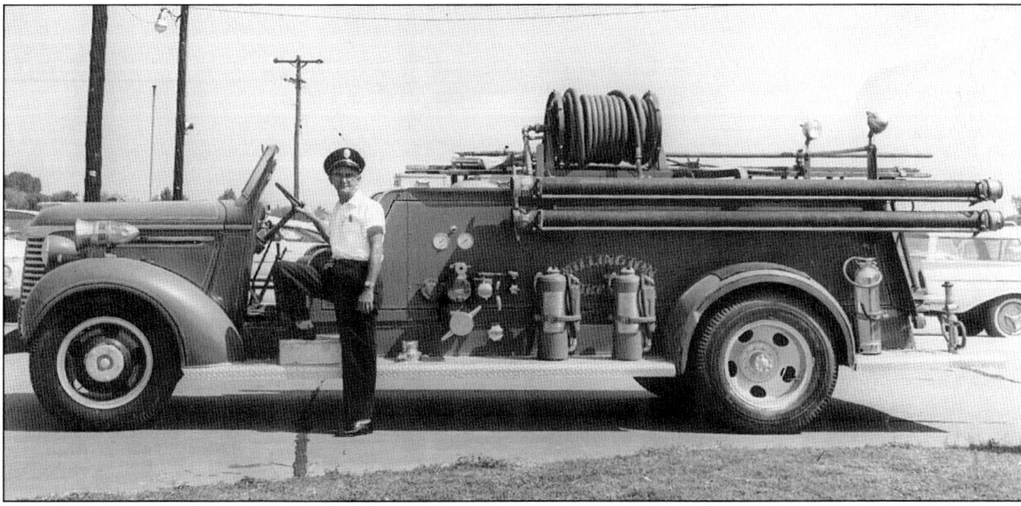

A volunteer bucket brigade narrowly averted disaster in 1902 when the home of D.G. Jennings caught fire and flames threatened to destroy the town. Sixty years later, Ed Childress (above) and other members of the Millington Fire Department were glad to have modern firefighting equipment when emergencies occurred. Below, James "Buddy" Oakley climbs into another engine outside one of the town's four fire stations. More recently, the fire department has added rescue operations and aircraft rescue firefighting to its services. Over the years, Millington firefighters have hosted and participated in activities for the community, including tours and demonstrations for school and Scout groups and even a haunted house at Halloween.

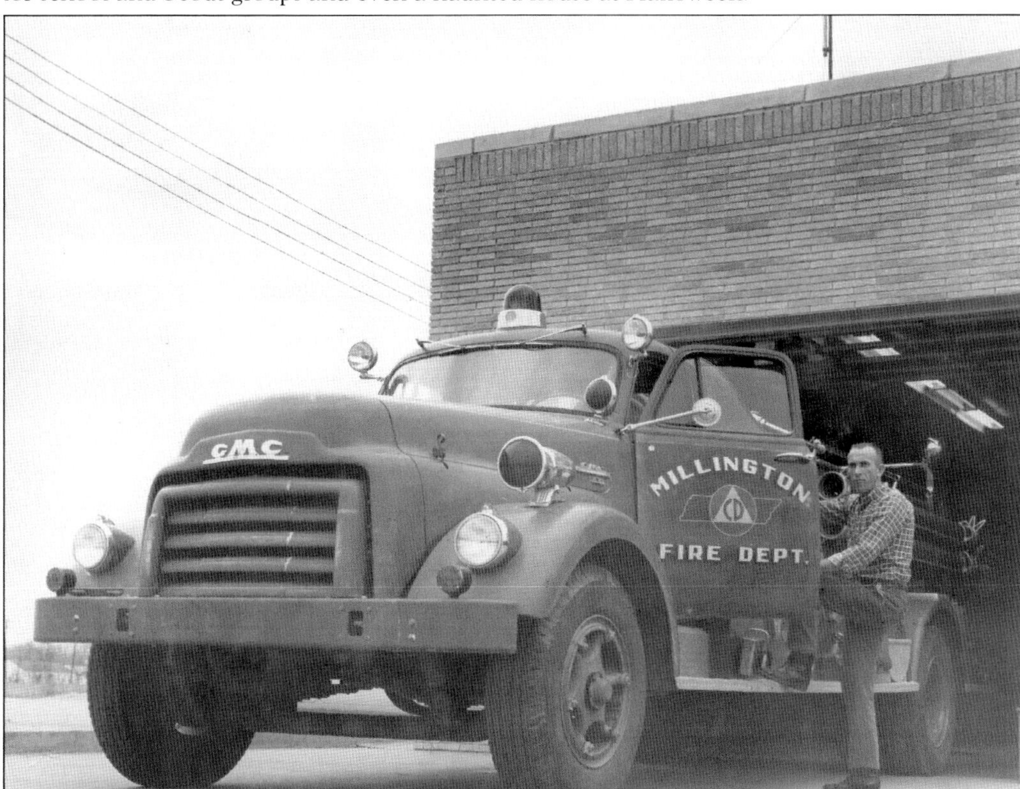

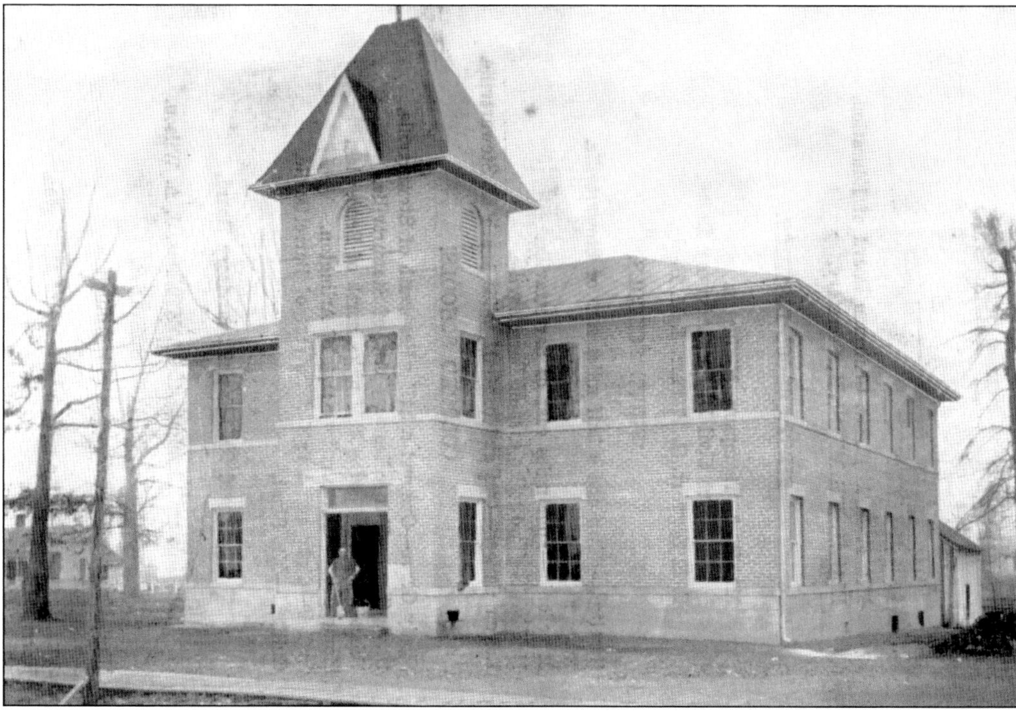

Until the 1920s, the one-room schoolhouse was typical and offered instruction to children of all ages in the same classroom, usually only through the equivalent of the eighth grade. Often, teachers were young people who had only recently completed the school themselves. This began to change in the early 20th century with school consolidations and the addition of secondary classes taught by college-educated faculty. Millington's one-room frame school was replaced in 1907 with the brick structure seen above. The smaller building on the far left was the home of the principal and has since been moved across the street. It has been the home of former principal William Osteen and his wife, former teacher Faye Ellis Osteen, for many years. By 1922, another new building was in place, as seen below. (Above, courtesy MPL; below, courtesy SCA.)

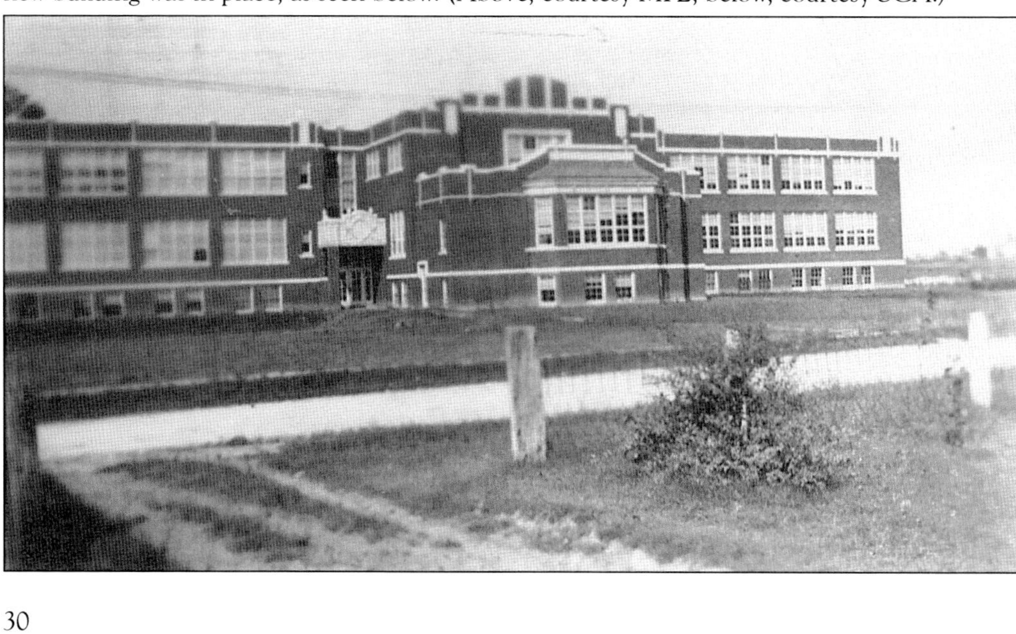

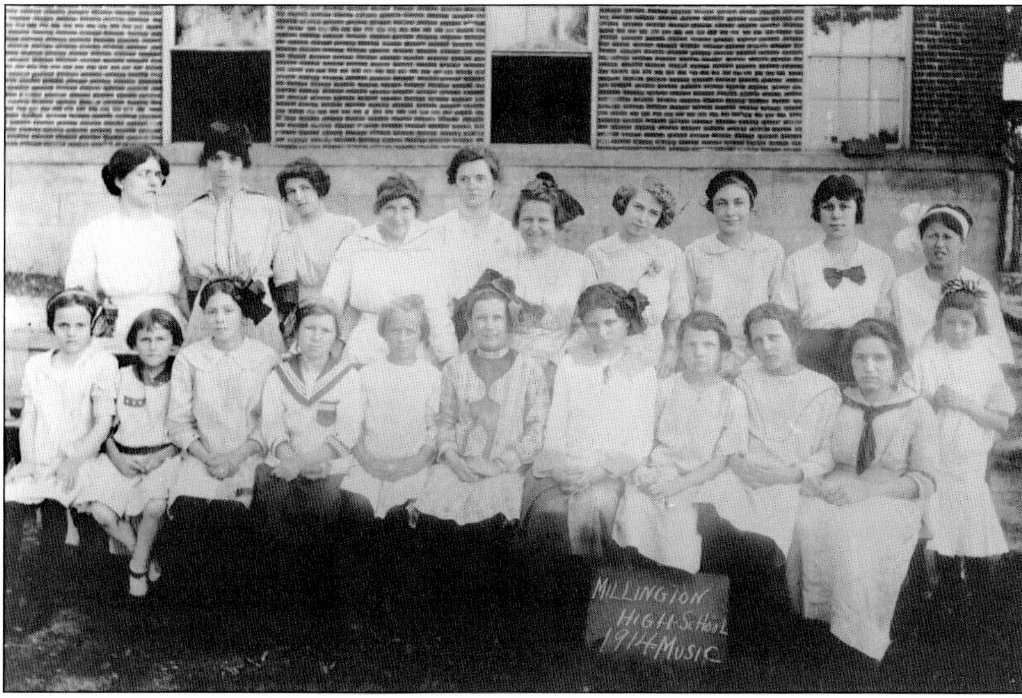

An interest in cultural edification permeated the county prior to 1920. Millington had one of the first schools in the county to add a music class to its curriculum for young ladies. The 1914 music class is pictured above. Millington Central High School's new gymnasium, completed in 1934, replaced the county's first school gymnasium, built in 1922. Eleanor Castles, director for women and girls' activities, expected it to "become the center of communities for both child and adult." Castles, supervisor of men and boys' activities Herman Osteen, and football coach W.D. Croft coordinated activities for the community center. (Above, courtesy MTC.)

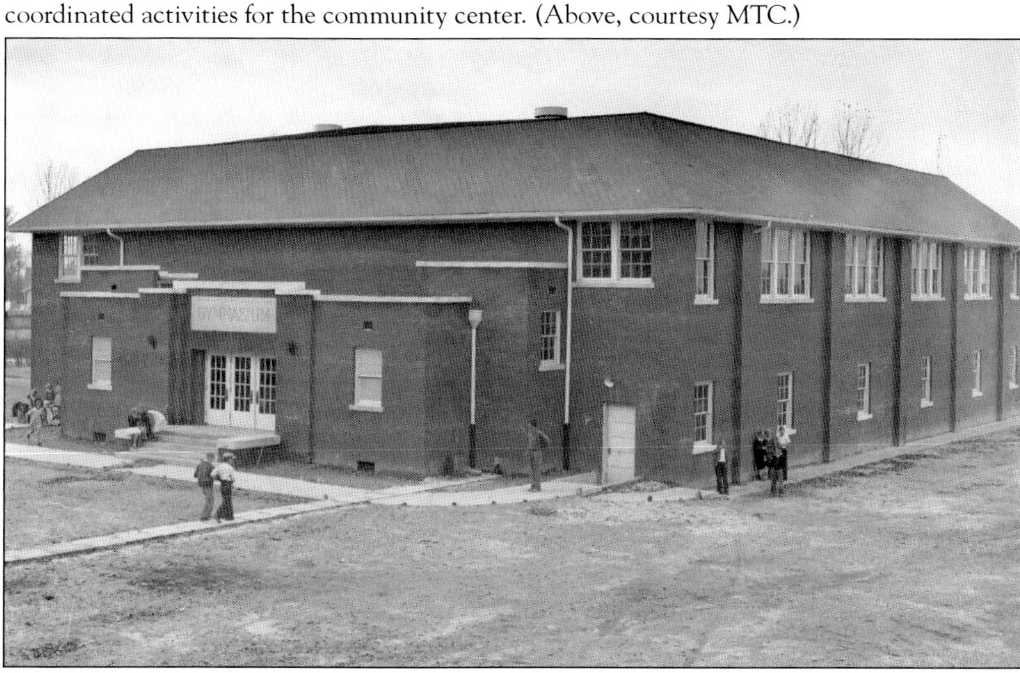

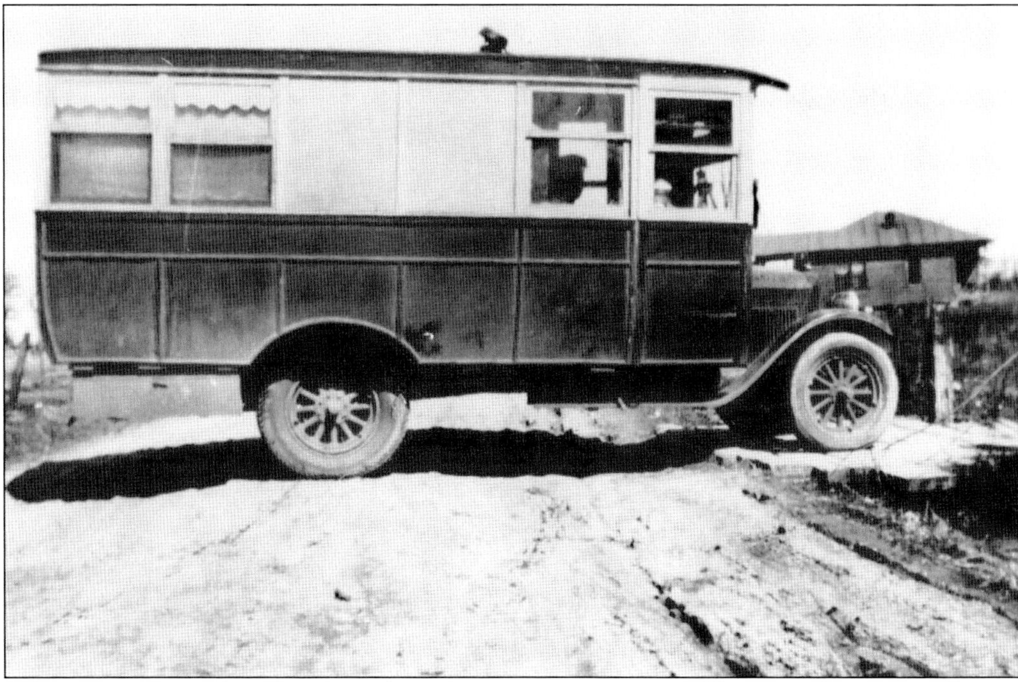

Millington students were some of the first in the county to benefit from transportation to and from school. The county school board established a wagonette system in 1907 that used mules to pull covered wagons with benches that ran the length of the bed. Motorized buses like the one above replaced wagonettes but kept the curtained windows and the option of providing heat on cold winter days by way of a stove in the back of the bus. Rapid evolution in the automotive industry soon replaced these early buses with more modern ones, like the ones pictured below, whose drivers wait patiently for the final bell of the day. (Both, courtesy MTC.)

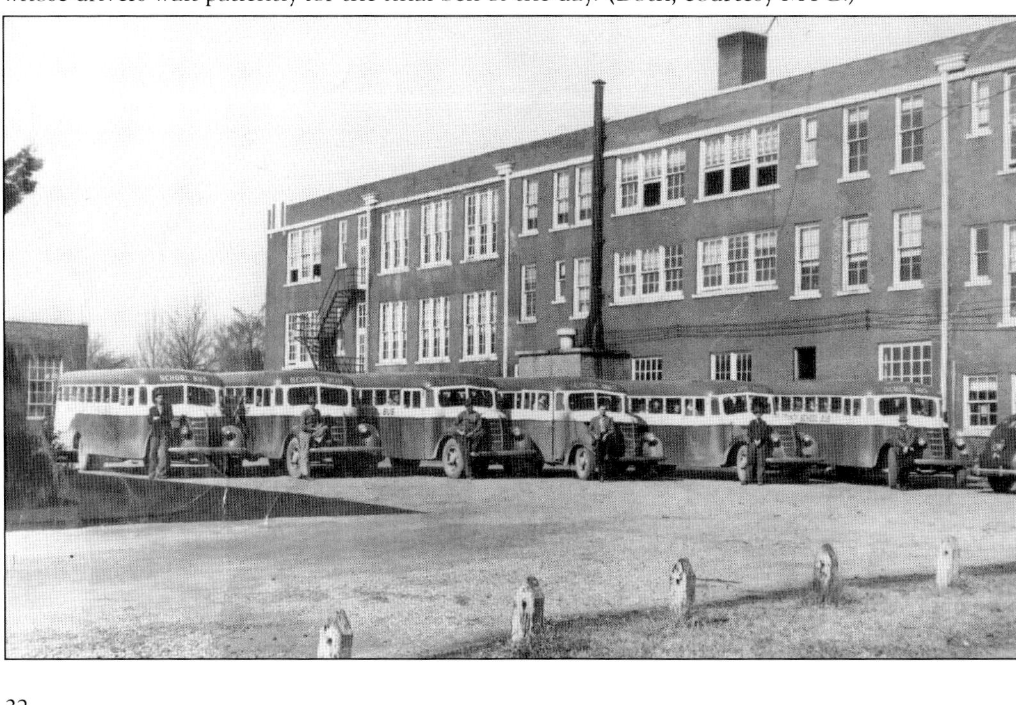

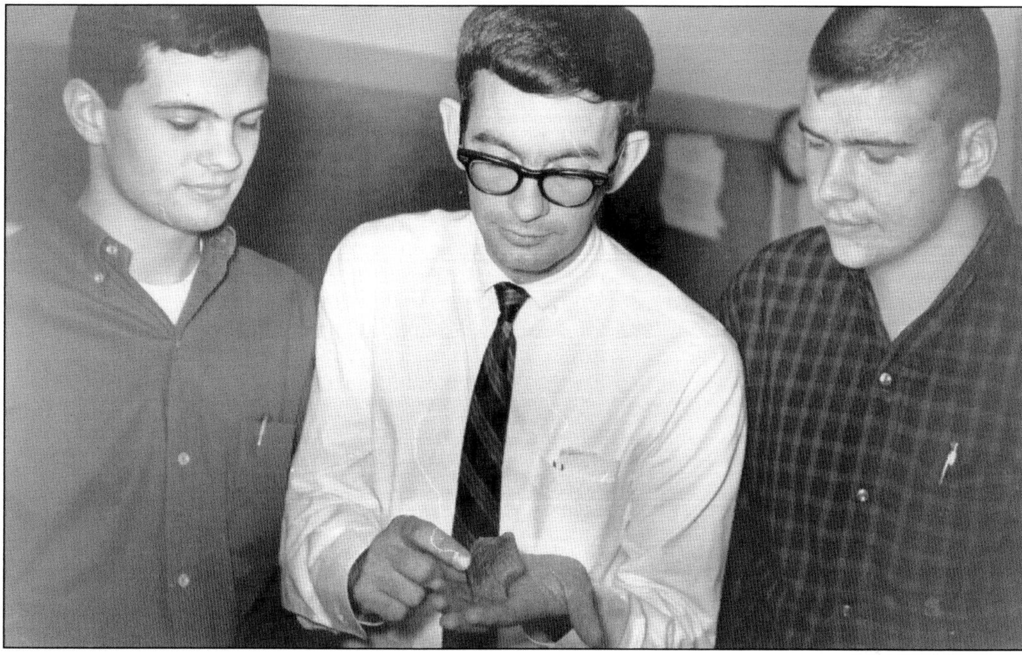

Millington Central High School was one of the first schools in the county to offer an archeology club, founded in 1963 by faculty advisor Aden Field, pictured above (center) with students Fred Prouty (left) and Mike Armour (right). The mission of the club was to foster an appreciation of the bluff area's prehistoric past and share that knowledge with archeologists at then-Memphis State University's Chucalissa Indian Village in southwest Memphis. One of the Millington area's earliest settlers, Clement McDaniel, is said to have conducted an active trade with local natives. In addition to the carved stone Prouty found, the club members in the photograph below (from left to right), Sherry Willis, Larry Murdock, Van Biggs, Bob Critchett, and Teresa Wade, unearthed a number of artifacts, which they proudly display.

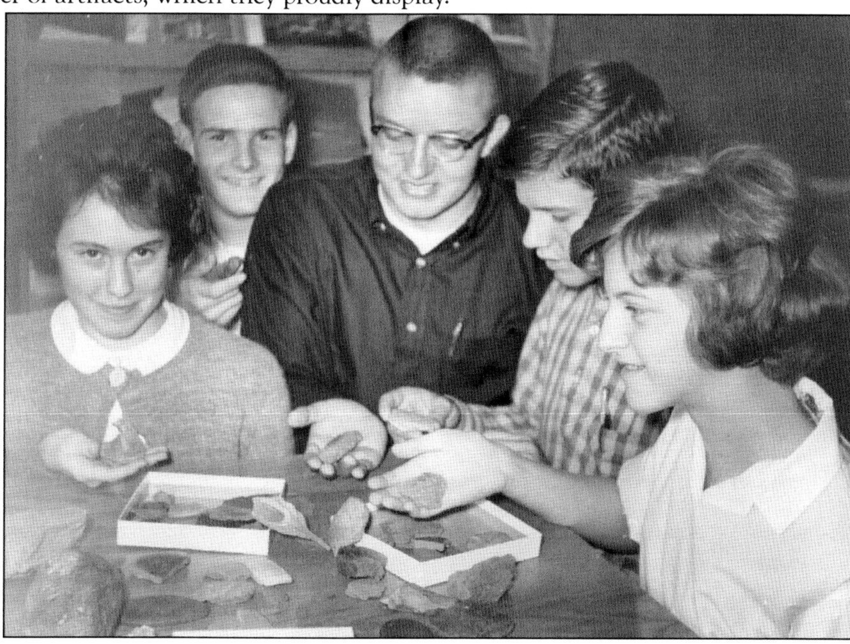

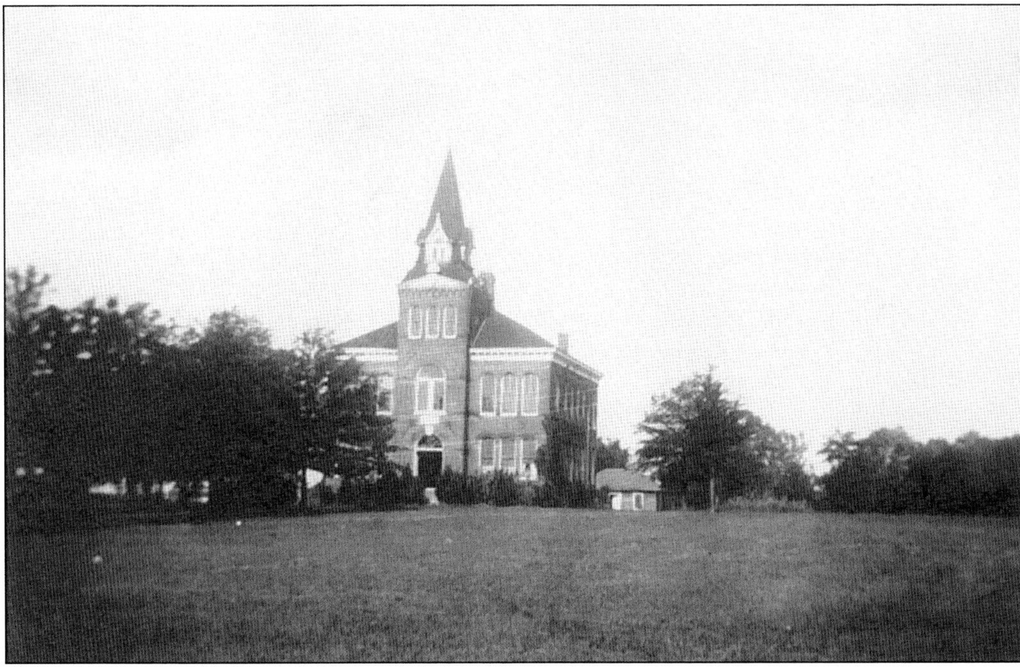

Several other schools surrounded Millington, at least until consolidation sent all area high school students to Millington Central High School. When Wade Bolton died in 1869, his will provided 1,200 acres of his Hoboken Plantation for use as an agricultural college, making this Tennessee's only secondary school situated on a land grant. Known as Bolton College from 1887 to 1922, the agriculture and teacher training school joined the county school system and, in 1925, became a secondary school now known as Bolton High School. There was also an elementary school on the grounds at one time. Pictured below in May 1964, Bolton students enjoy field day activities. (Above, courtesy SCA.)

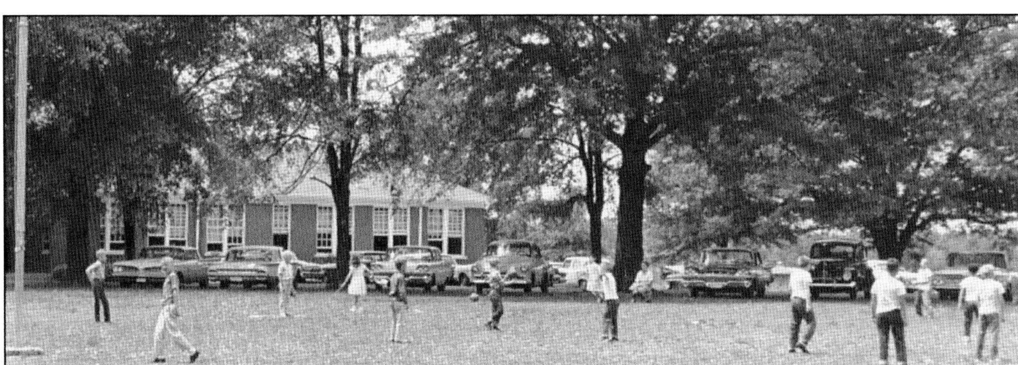

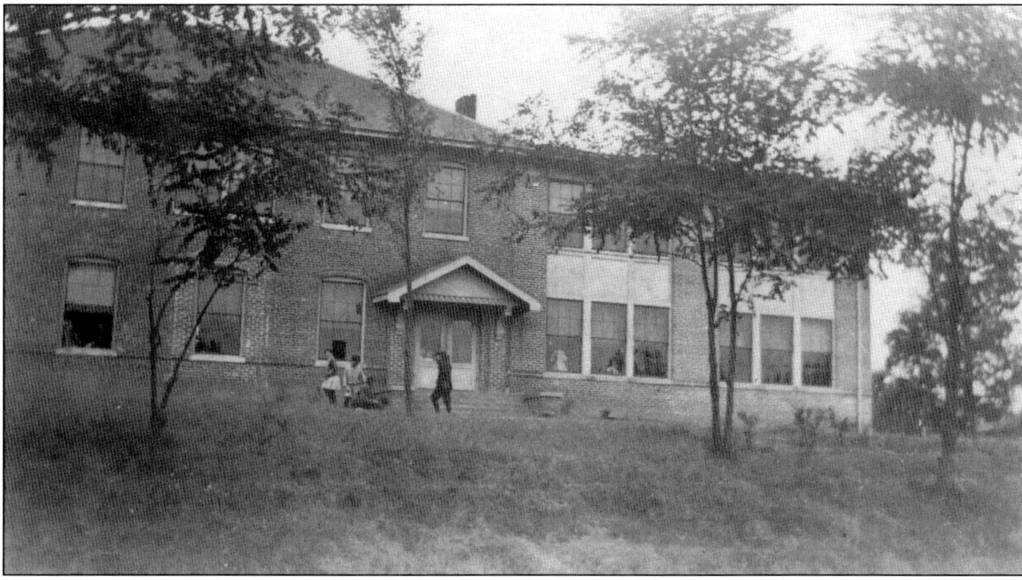

West of Millington, Cuba High School replaced Shady Grove School in 1907, and in 1909, a two-story brick building with nine rooms and 275 desks was built next to the Masonic Lodge. It included an auditorium and library, and in 1916, an addition housed the domestic science department's four oil stoves and sewing machines. Shown here in 1922, Cuba School burned during the winter of 1923. Classes were held through the spring term in the Masonic Lodge, and beginning in fall 1923, students were bussed to Millington. Pictured below is the Cuba High class of 1914, which included first through 11th grades. Students wishing to complete the 12th grade often attended West Tennessee Normal School (now the University of Memphis), which offered 11th and 12th grades when it opened in 1912. (Above, courtesy SCA; below, courtesy MTC.)

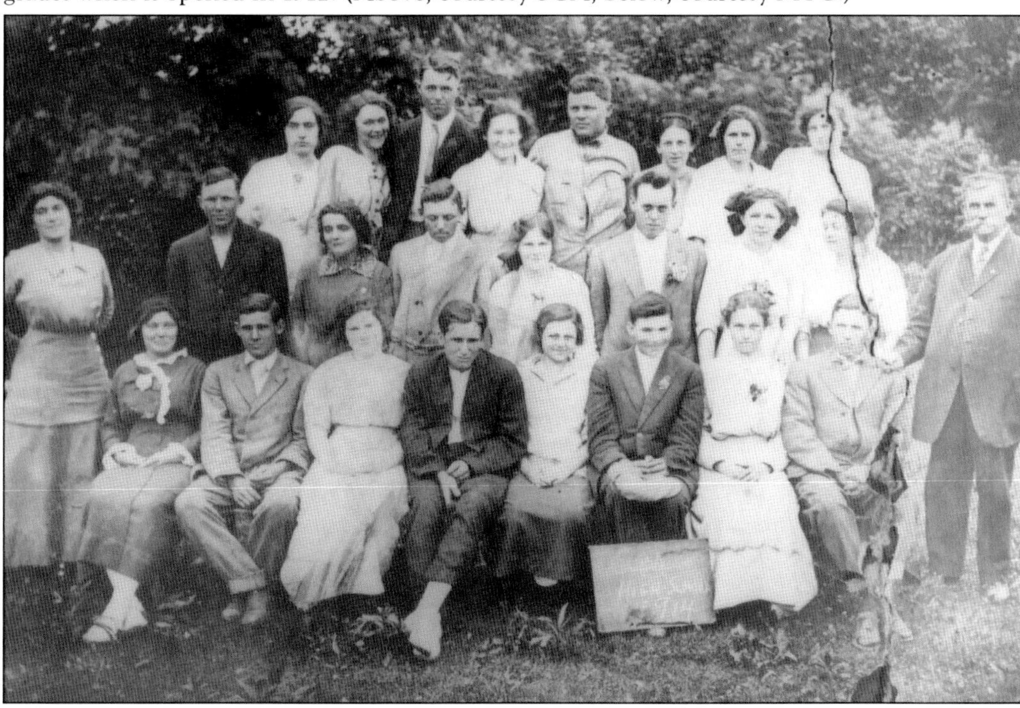

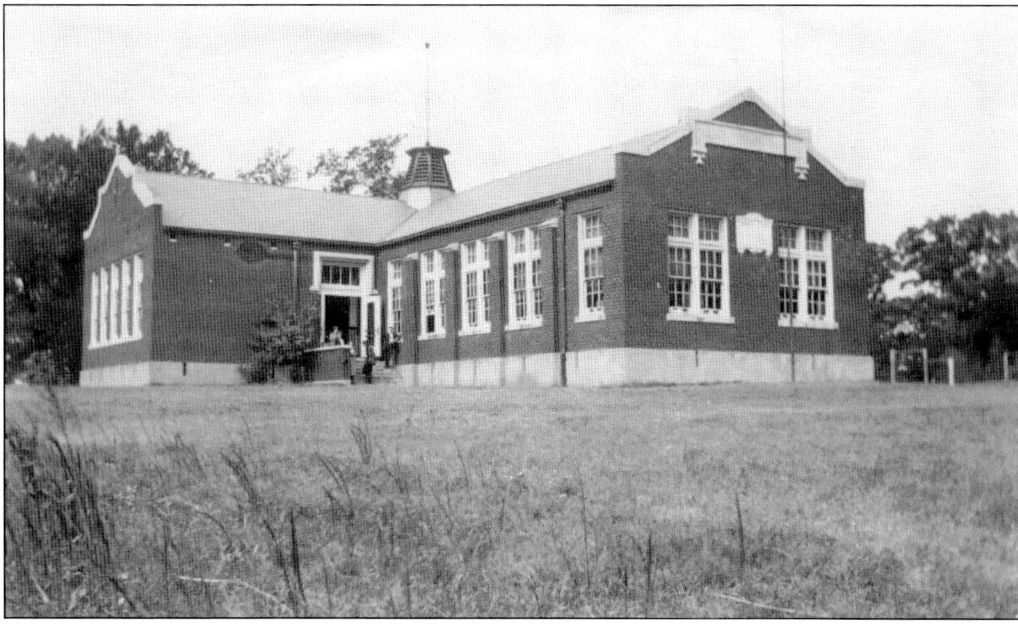

Lucy High School students came from Big Creek, Pleasant Ridge, Rembert Springs, and Woodstock grade schools. The school grew from a two-room building in 1875 to this much larger brick structure on property donated by Lewis Williams in 1910. In 1920, the high school students transferred to Millington, as did seventh and eighth graders beginning in 1927. A primary school only for the next 13 years, Lucy School again offered classes through eighth grade after 1940. Below, an early Lucy High School basketball team poses for a group photograph. (Above, courtesy SCA; below, courtesy MTC.)

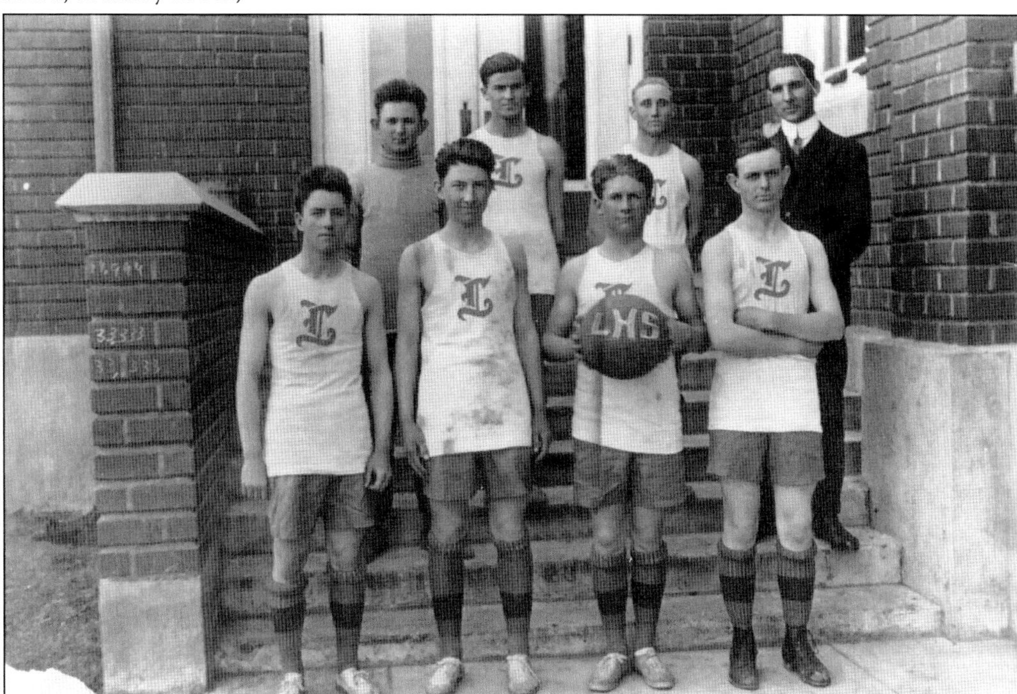

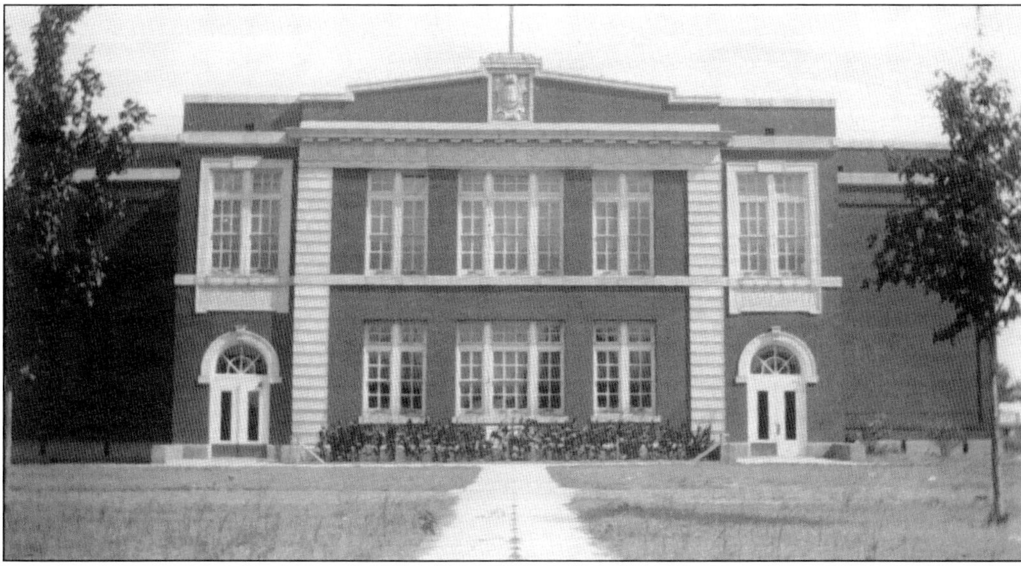

Rosemark School began as a grade school, became a high school until 1930, and was again an elementary school from 1930 to 1967 when the elementary school closed. In 1970, the county sold the property to the private elementary Tipton Academy at Brighton, which renamed it Rosemark Academy and used the building for high school classes. In 1988, the two schools merged, becoming Tipton-Rosemark Academy. The original building was demolished in 2008 amidst much protest from preservation-minded citizens. (Courtesy SCA.)

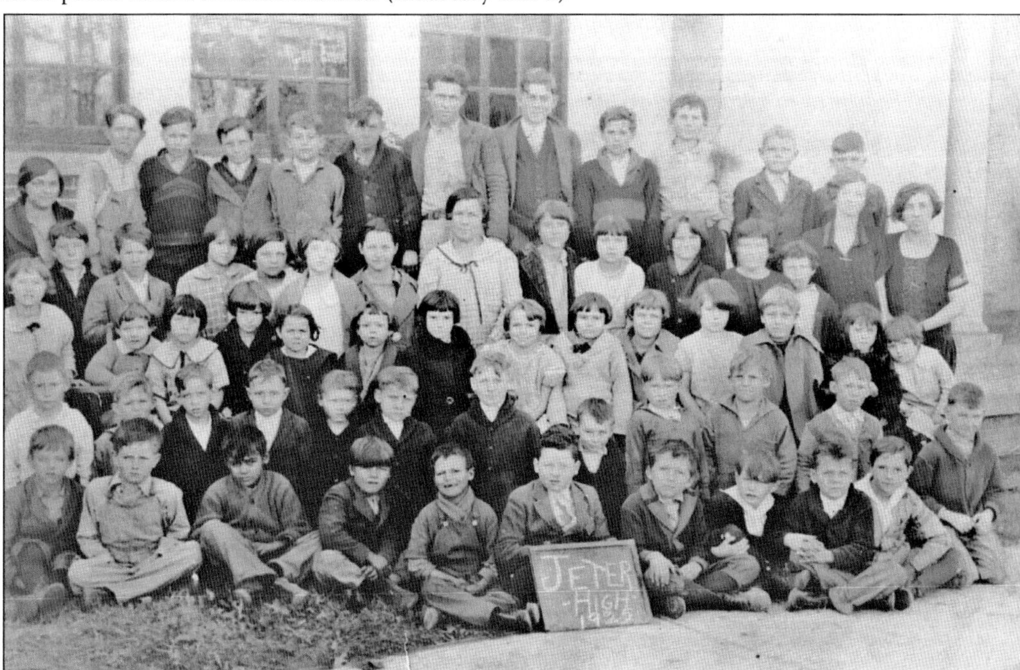

Students from first through 11th grades comprised the 1925 class of E.E. Jeter High School. Named for Squire Emmet Early Jeter, the school has served primarily as an elementary school since the late 1920s. A new building was constructed in 1970 on the original Benjestown Road site. (Courtesy Doug Ammons, Shelby Forest General Store.)

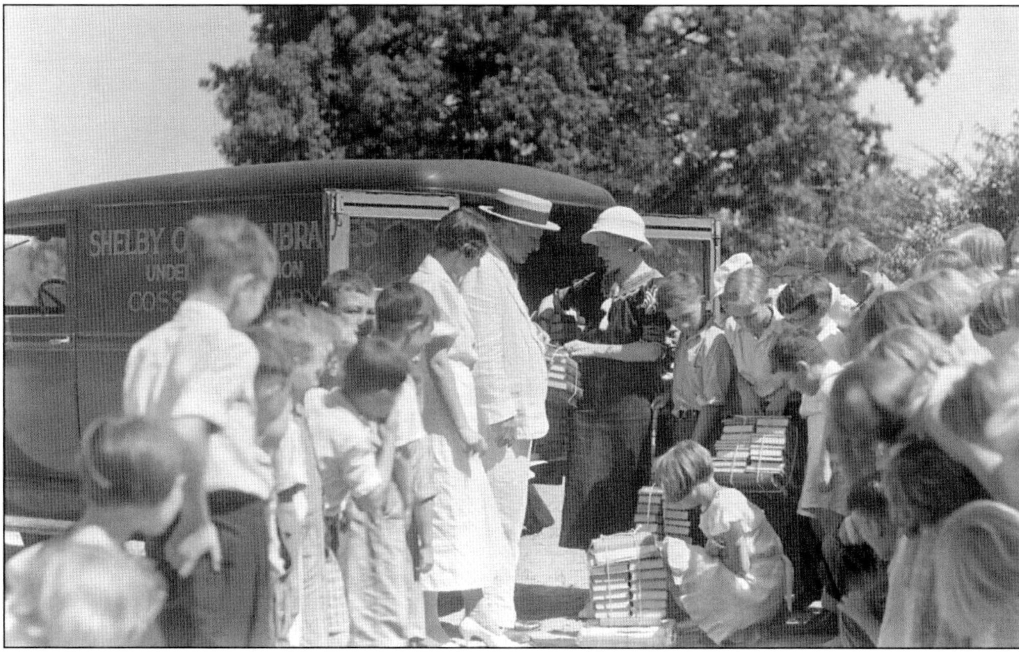

A mobile library system was established by the county court in 1907 at a cost of $200, which provided 30 boxes of 10 new books to rotate through the schools each year. More books were added to the rotation over time, until permanent libraries could be established within each school. Miss Miller, the county's first bookmobile librarian, delivers books to the students of E.E. Jeter School. (Courtesy MPL.)

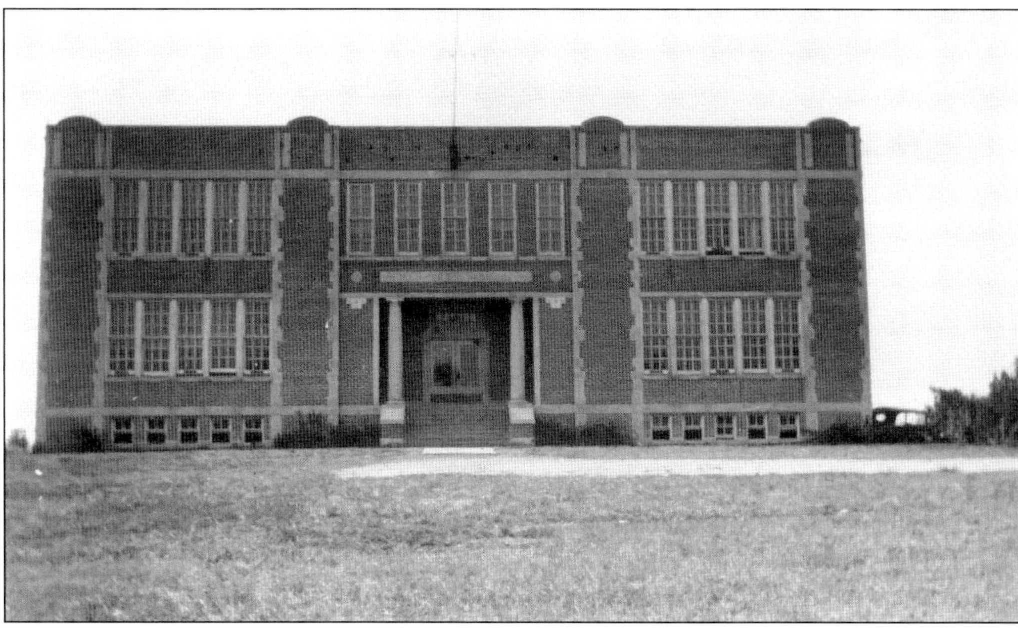

Kerrville School was established in the 1870s and included a boardinghouse for teachers, who were not allowed to marry by state law. A two-story wooden schoolhouse was destroyed in 1913 by a tornado and replaced by this brick building. When Kerrville School closed and was demolished in the 1930s, its students transferred to Millington. (Courtesy SCA.)

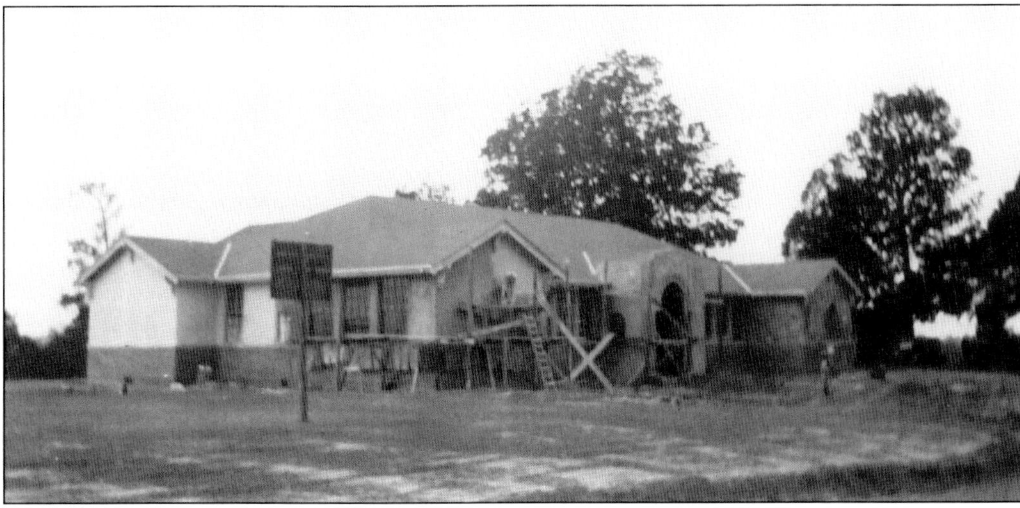

Locke School, shown here under construction in 1922, replaced the one-room Toof Academy and offered grades one through 11. The Locke community was named for the family of George Washington Locke after an application for a post office to be named Giles Town was denied. The Locke and Giles families were related by marriage. (Courtesy SCA.)

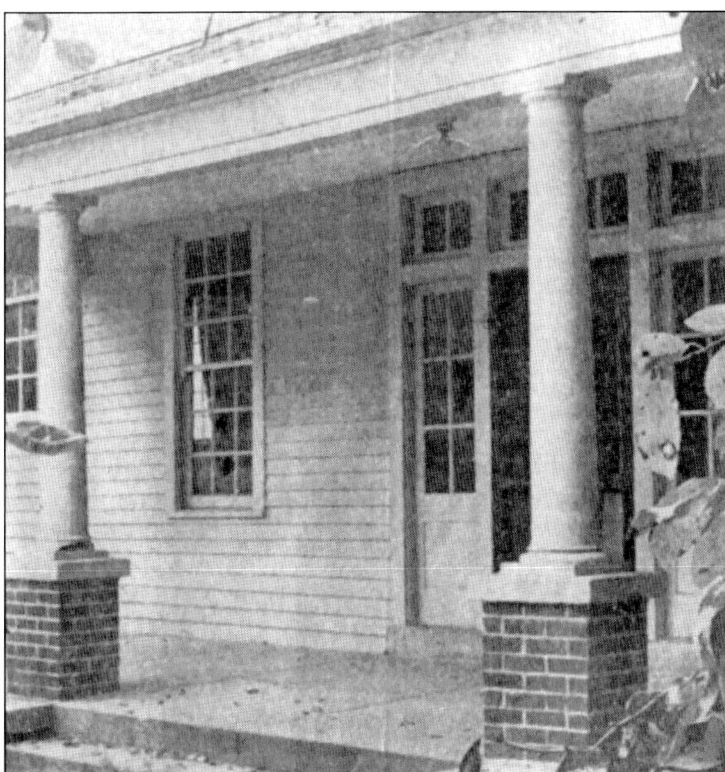

In the 1890s, anti-lynching activist Ida B. Wells taught in the Woodstock area. Shelby County Training School was a boarding school established in 1913 and the first to provide two years of high school to African Americans in the county. The school's mission was to teach farm and home skills to rural students. A dormitory, pictured here in 1972, was added before 1920. The school was renamed Woodstock High School in 1963. In 1970, high school students transferred to Millington Central High School, and Woodstock became an elementary school.

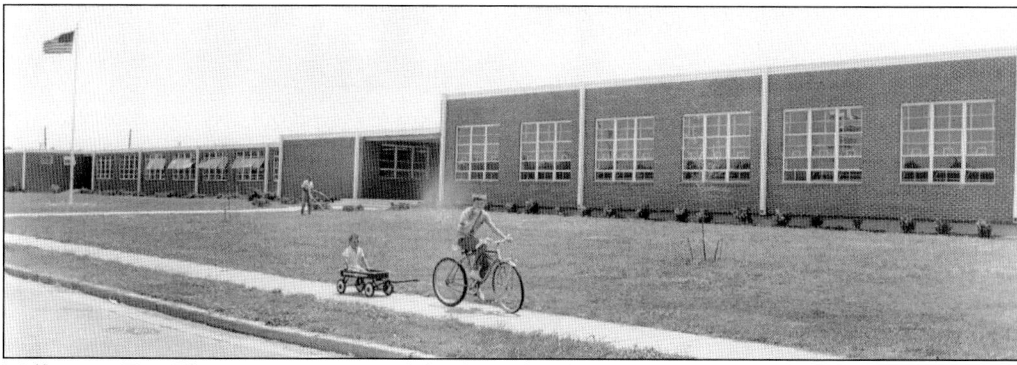

Millington East Elementary was one of the first schools desegregated in Shelby County. A federal request to assign all elementary students living on the Navy base to the new school was approved in 1963. Until then, African Americans attended E.A. Harrold Elementary, which was north of Millington, rather than the new school near the base.

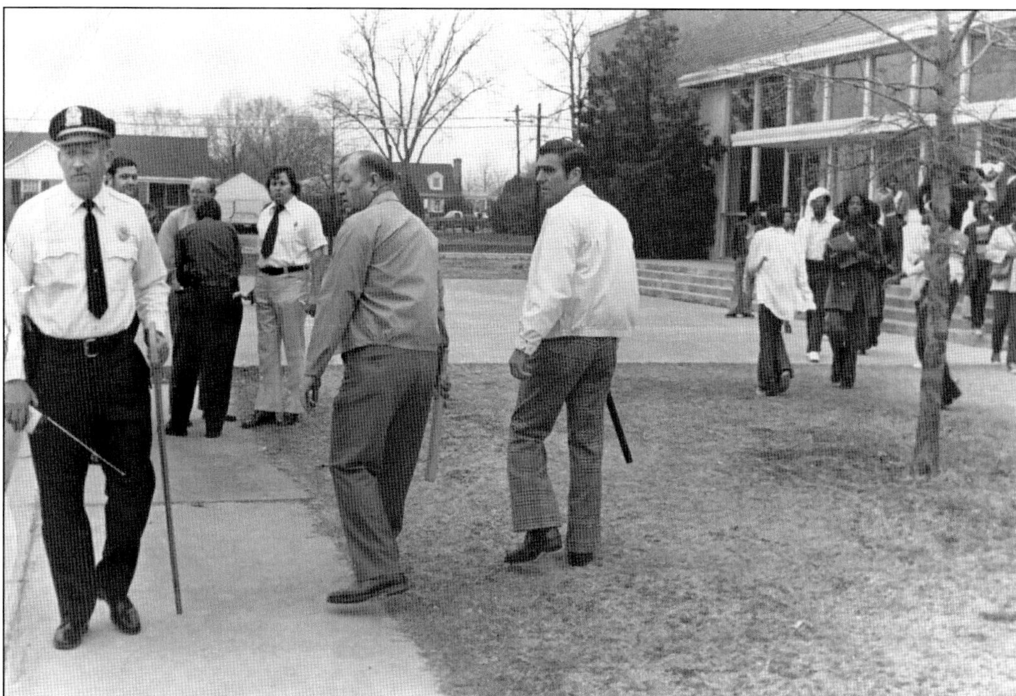

Desegregation was relatively smooth for Millington area schools with one reported exception. In 1974, some of the school's white students walked out of a Black History Week assembly, sparking resentment. The following week, African American students initially refused to attend an assembly, then arrived late and proceeded to the gym floor rather than the bleachers. Newspapers reported that fighting erupted when two or three white students jumped out of the bleachers and accosted them. Police were brought in to control the situation.

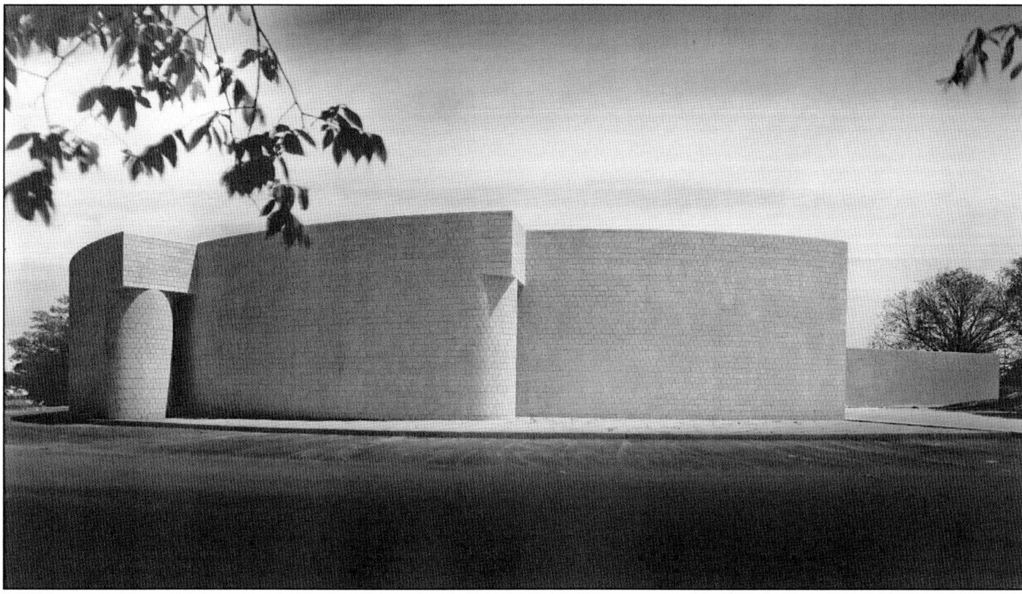

The then three-year-old Millington Middle School, above, won a national award in 1974 from the American Institute of Architects for its spacious, open-concept design, seen below. Several additions reflect the school's growth over the years. Concrete block partition walls were added to the interior in 1982 to achieve a more traditional environment. The partitions were remodeled in the 1990s, creating separate classrooms but destroying the school's original award-winning design.

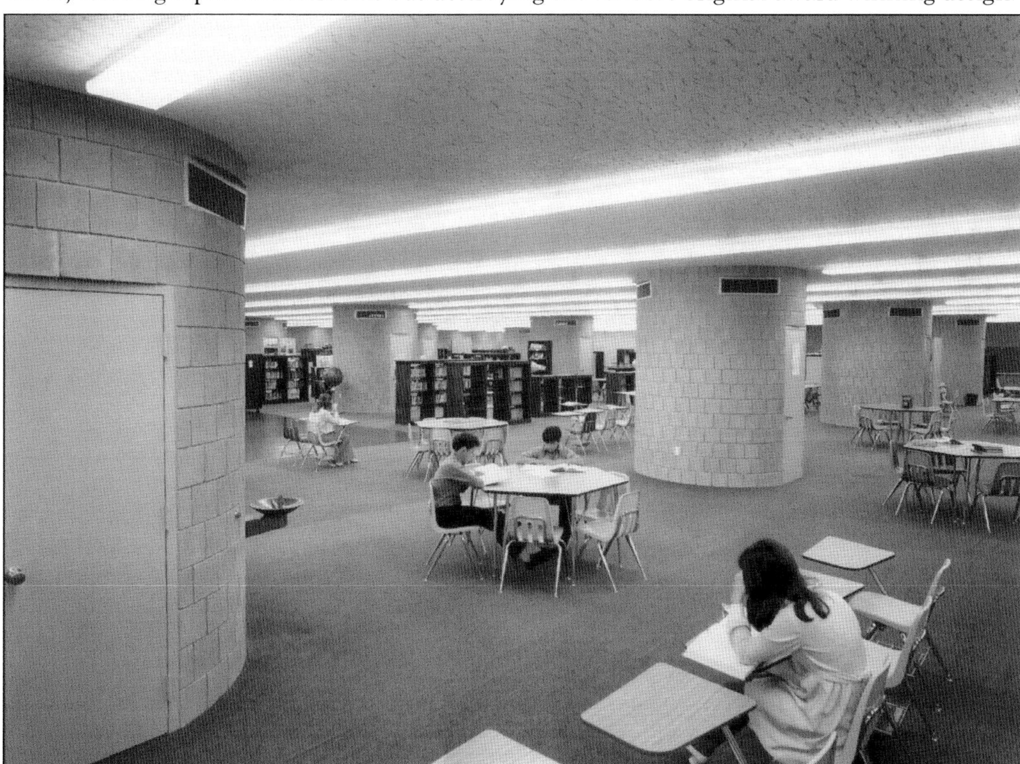

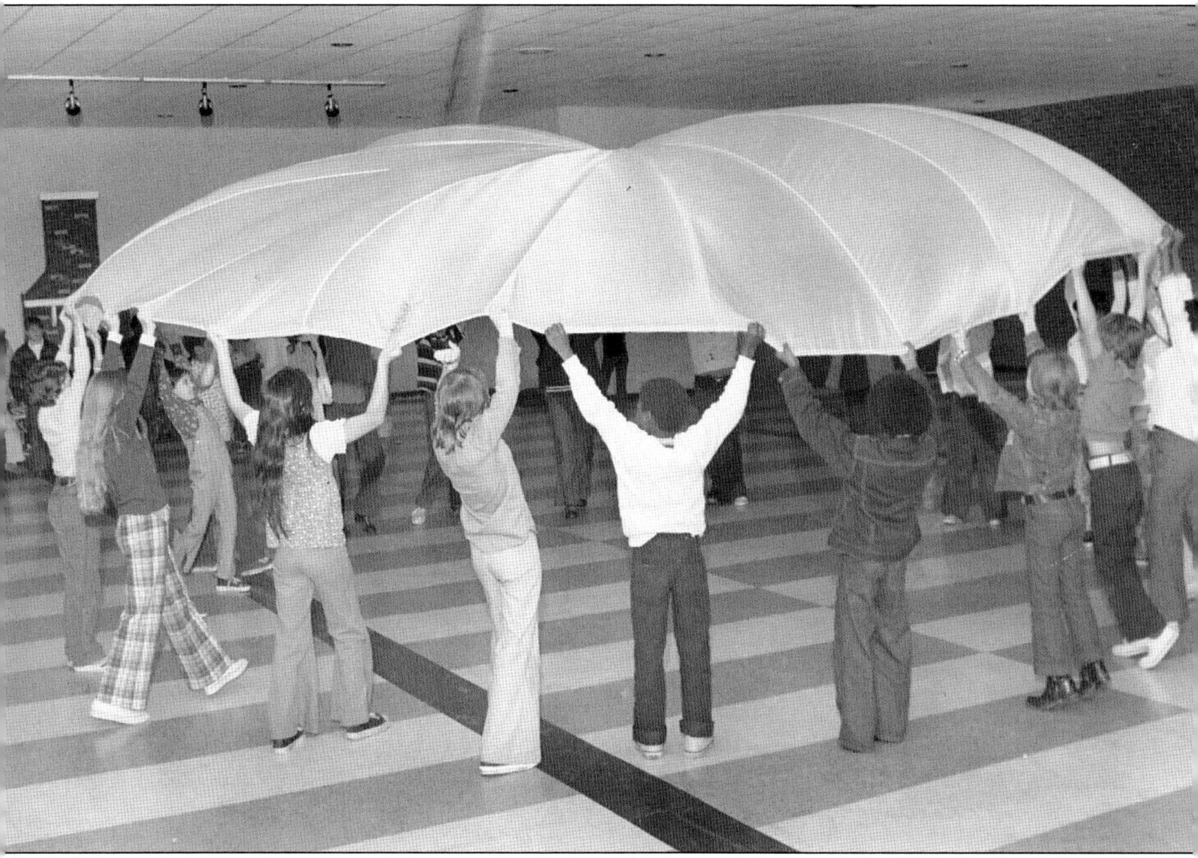

In 1975, Millington Central Elementary School students enjoyed a new activity in their physical education classes. Physical education instructor Eddie Harris was responsible for introducing a large parachute to the classes, which exercised group cooperation skills as well as hand and arm muscles, as students worked to keep it aloft. The school cafeteria was the only area with a ceiling high enough for this activity. The elementary schools consolidated in 2004 in the new Millington Elementary School on William Osteen Drive near the naval base, behind the old Navy hospital that now houses the University of Memphis's Millington campus.

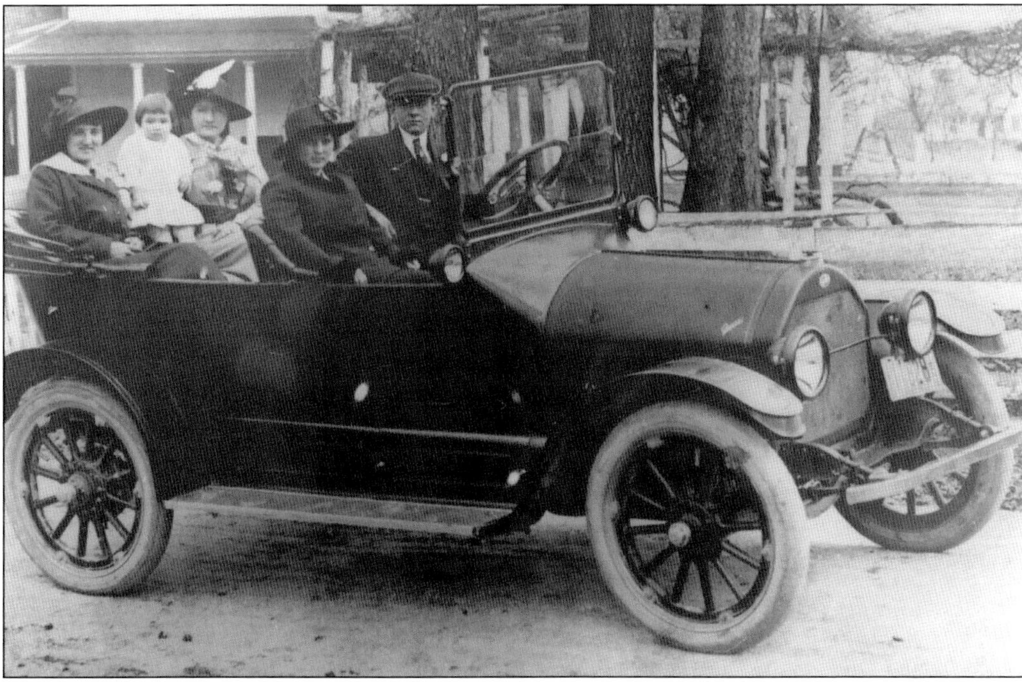

The first road order issued by the county court was to a settlement on the Hatchie River, far north of Millington. The second was for a road from Memphis to Big Creek. Little more than a wide dirt path took travelers home to Big Creek, but a plank road leading to Memphis helped prevent wagons laden with goods from sinking into the mud after a rain. Road improvements became increasingly important after citizens began purchasing automobiles. As pictured in front of the Pryor house in 1919, male chaperones often accompanied ladies such as Birdie Tucker (above, in front seat) and her companions (from left to right in back seat) Dimple Crenshaw, baby Velma, and her mother, Florence White, on trips to town for both safety and propriety. Below, a group of men pose around a car outside the E.A. Harrold Store. (Both, courtesy MTC.)

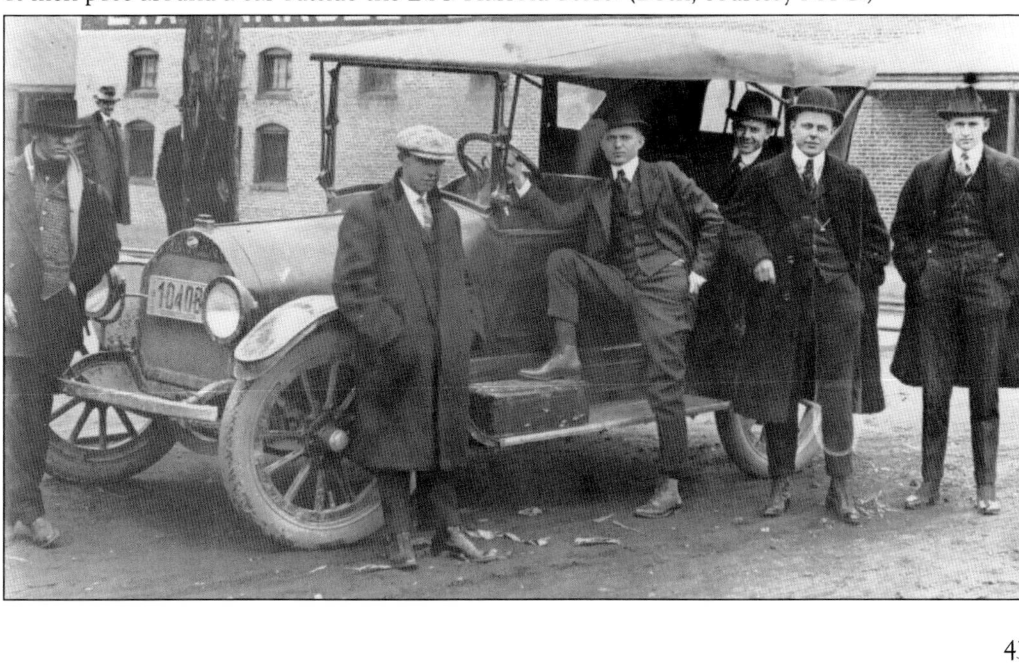

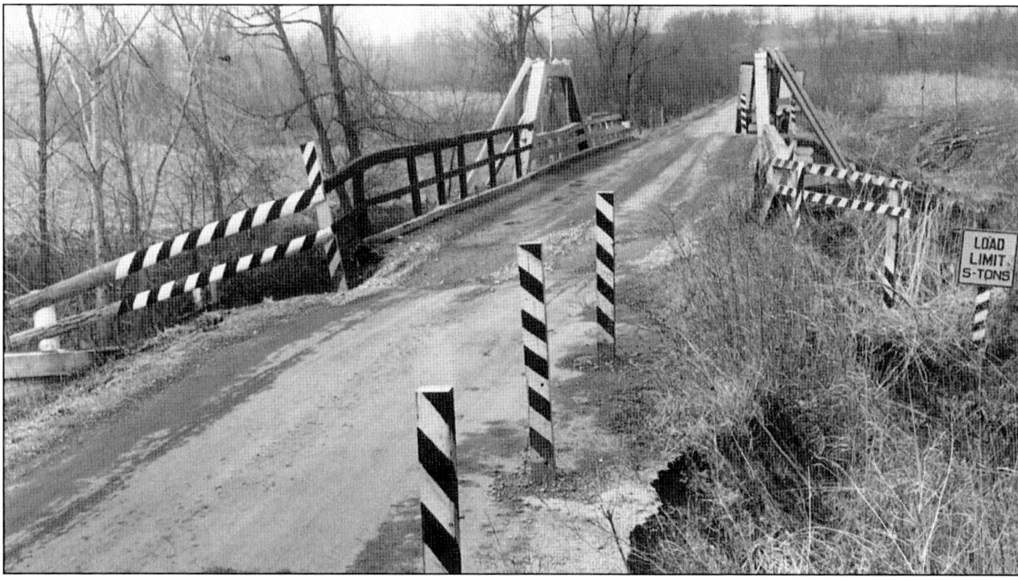

Built in 1916, the Mudville Road bridge over Big Creek was crooked, one-way, and, as noted by the *Memphis Press-Scimitar,* "unusually high" at 26 feet above the creek bed. A straight, two-way concrete bridge 600 feet south replaced the unsafe bridge in 1951. In 1955, an area resident semi-successfully petitioned the county to have Mudville Road renamed something "more pleasant and appropriate." The road remains Mudville, but the community it serves was renamed Pleasant View.

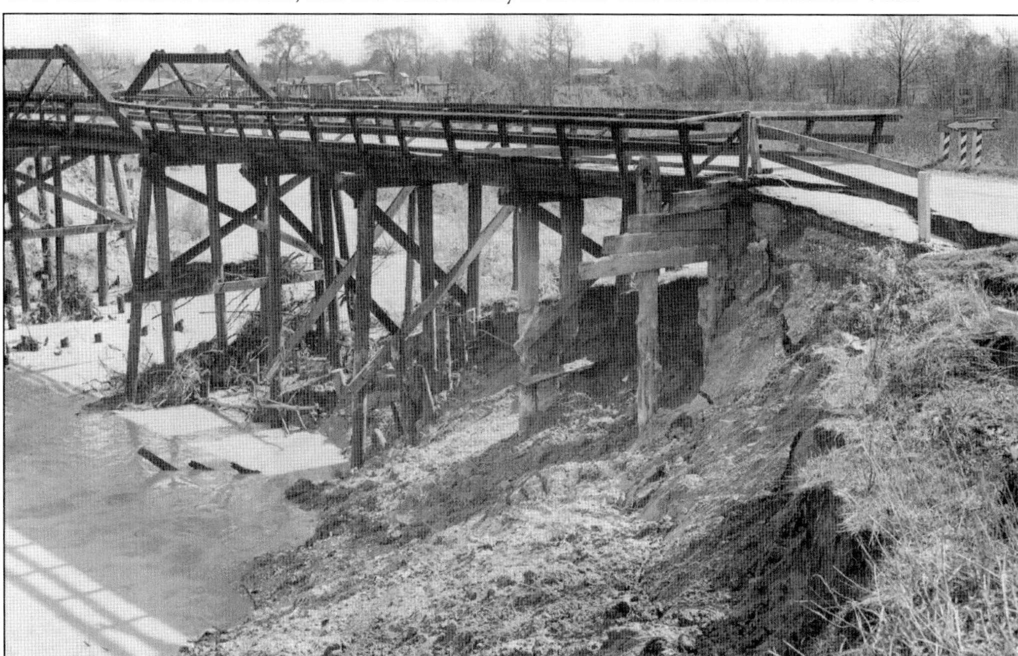

In 1949, trees and debris carried by floodwaters damaged the bridge over Big Creek on Raleigh-Millington Road, which was used heavily for access to the naval base. Big Creek frequently climbed out of its bed and wreaked havoc throughout the area until the Army Corps of Engineers created a drainage canal. The road was moved 900 feet east to meet a new bridge built to replace this one.

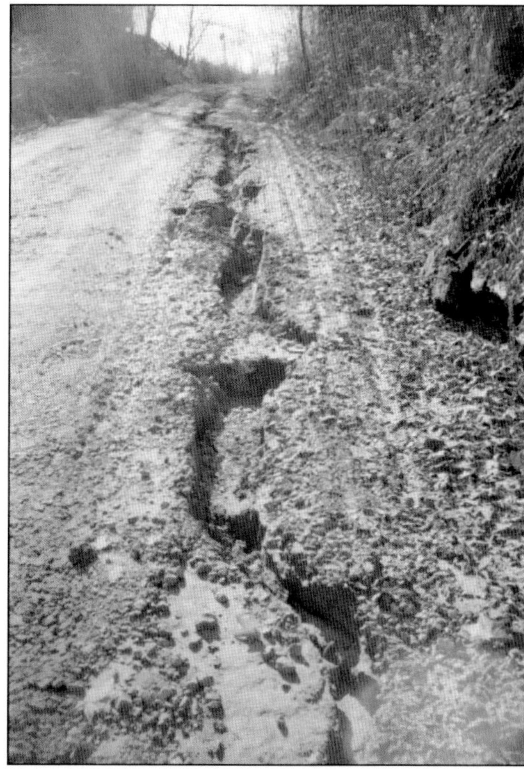

Many roads around Millington were in terrible condition by the 1950s. Beginning in 1949, Gov. Gordon Browning's rural roads project, designed to make transporting goods from farm to market easier, prompted a spate of improvements throughout Tennessee. Herring Hill Road (right), west of Millington, was so badly rutted and prone to rapid run-off during rain, it forced some residents to leave home at the first sign of a storm to avoid being trapped in a quagmire. Others were unable to find boarders, despite a housing shortage in the area, due to the potential for damaged vehicles. Below, W.S. Rast Sr. and William Geno spent November 21, 1952, draining wallow on Herring Hill Road in preparation for Thanksgiving visitors.

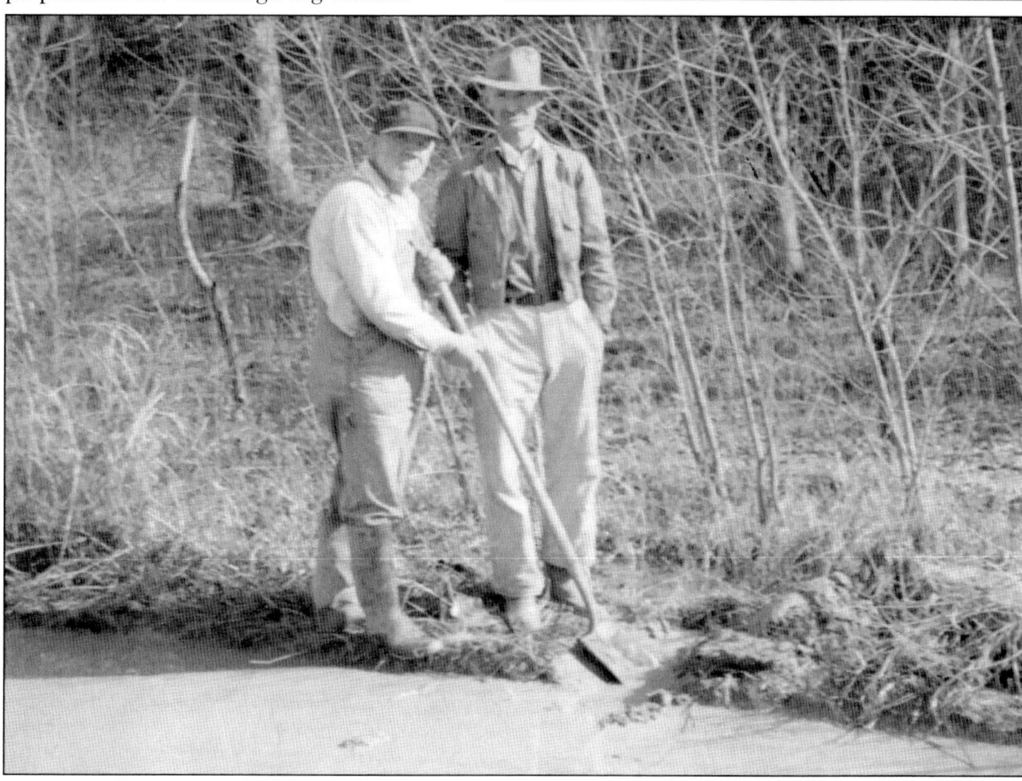

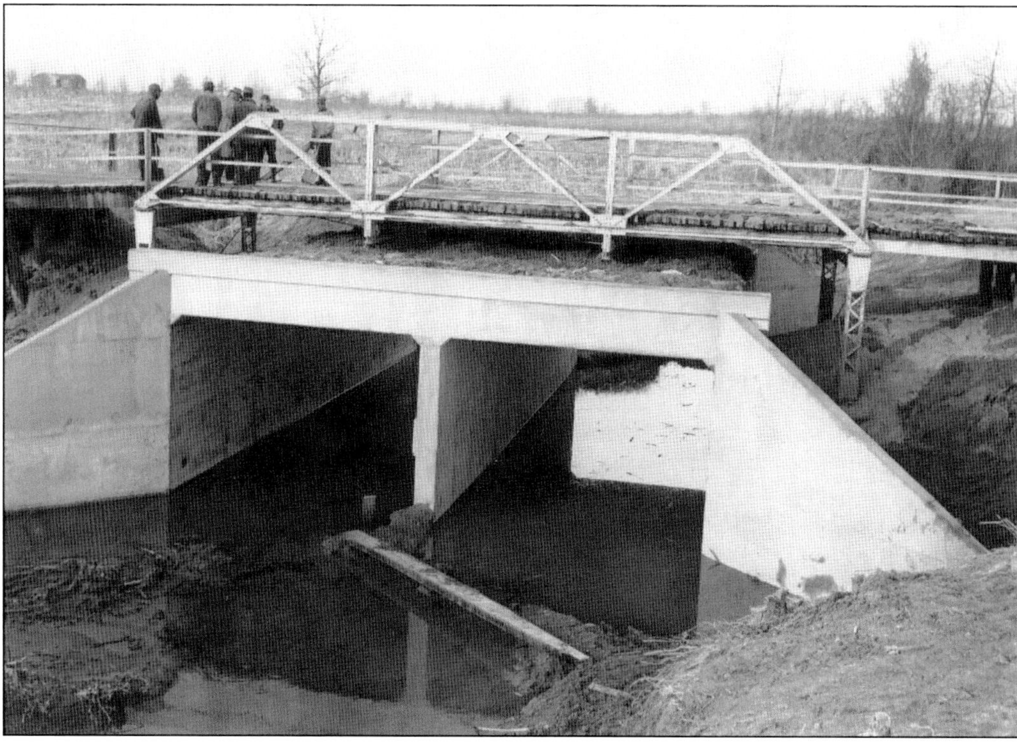

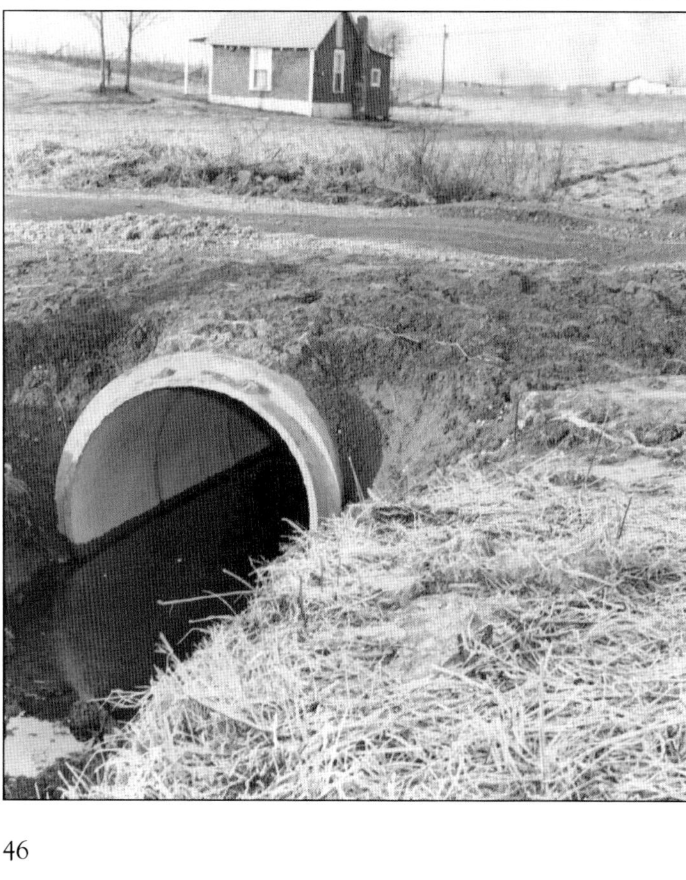

In 1911, Memphis Bridge Company erected a steel bridge, seen above, on Woodstock-Cuba Road over one of the many creeks in north Shelby County. Forty years later, the county replaced it with a concrete box culvert underneath the old steel bridge that was later removed. Woodstock-Cuba Road is one of the access roads leading to Meeman-Shelby Forest State Park. At left, a one-way timber bridge along this road was replaced in 1951 with a concrete culvert as the county worked to improve drainage and prepared to upgrade the dirt and gravel roads in this area to asphalt. Efforts to improve drainage throughout the flood-prone region continued for more than another decade.

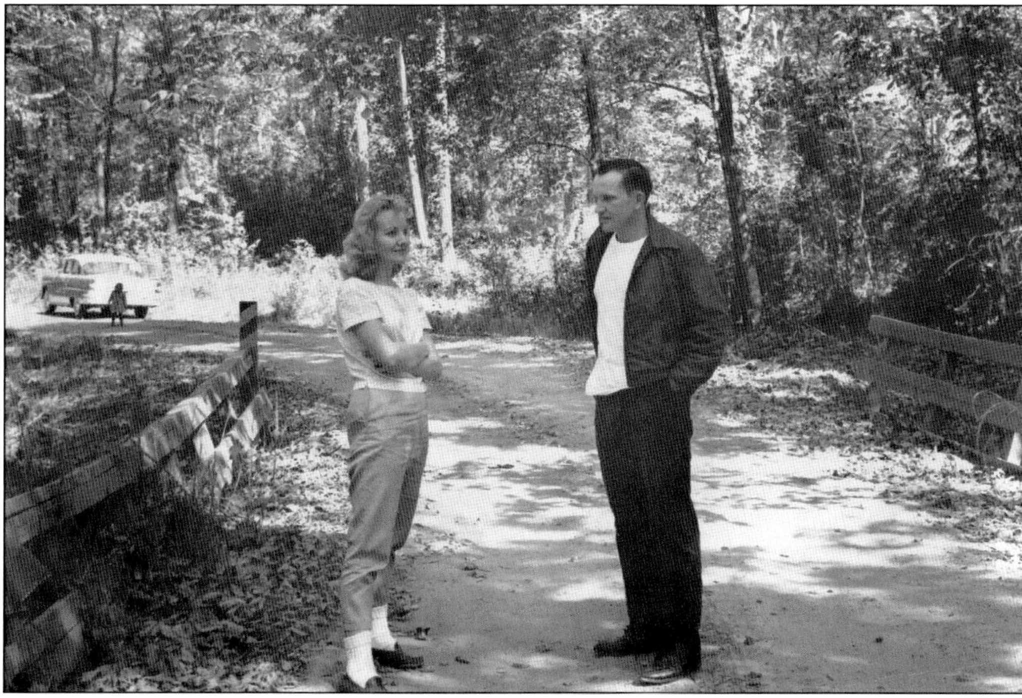

A steep road favored by area children for sledding during snowy winters, Herring Hill Road residents continued at least into the 1960s to deal with serious erosion, prompting neighbors Peggy Dicus and George Rast, above, to discuss a bridge that had partially collapsed. Although the county had long promised to remedy the situation, little more had been done to improve the road than the addition of gravel or oil to keep down dust. Dicus and Rast, as well as Lillian McClelland, the Cliftons, the Lucados, and other families along Herring Hill Road, wanted a solution to the drainage problems and the installation of a hardtop road like others in the area. Finally, in October 1961, workers from the county penal farm were assigned to dig ditches and prepare the road for resurfacing the following spring.

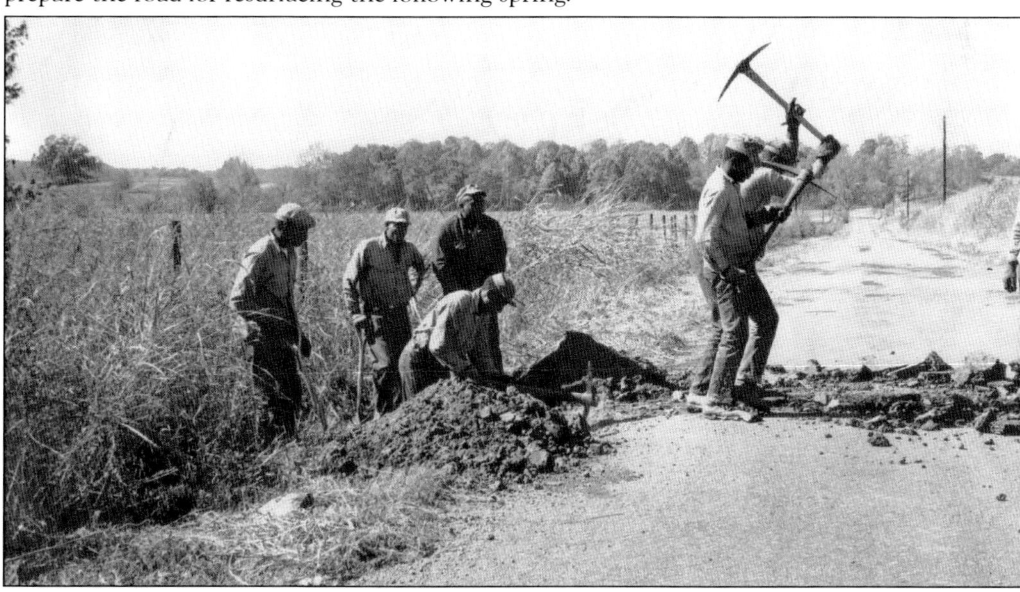

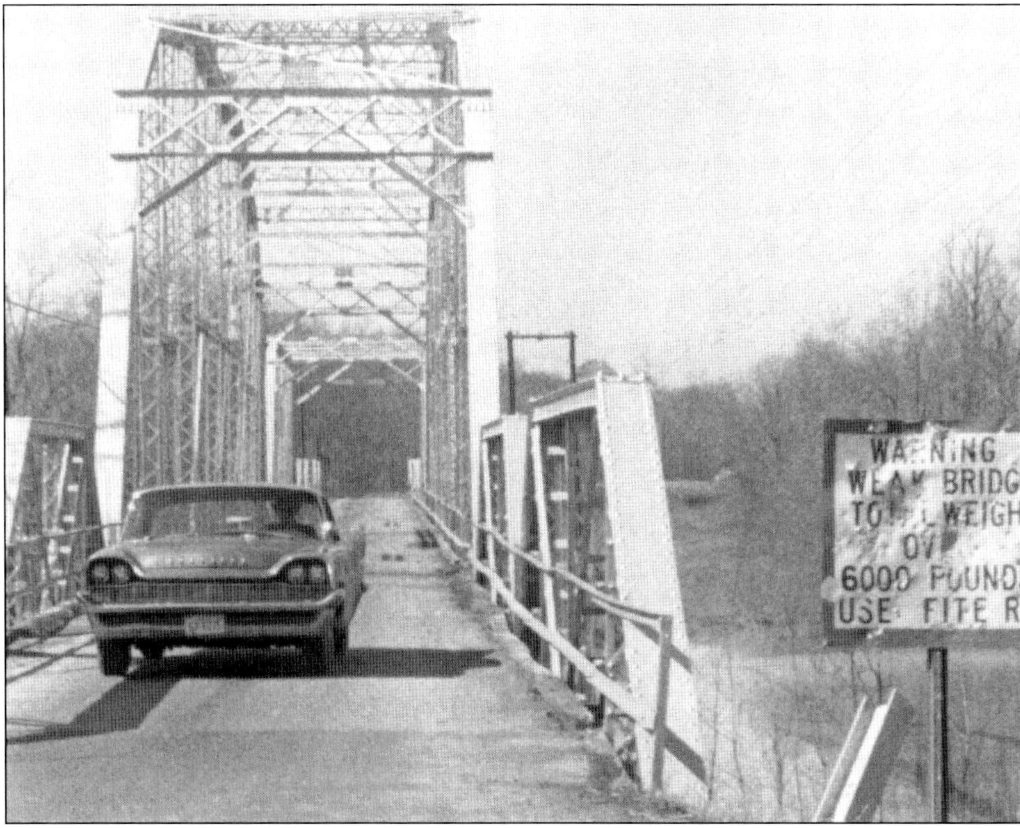

This steel bridge once stood on Benjestown Road as the only crossing over the Loosahatchie River west of Highway 51. Built in 1913, the rickety structure could accommodate only one car at a time. Pictured here in 1966, it was replaced the following year by the North Watkins Extended Bridge as part of a road-widening project designed to better serve new housing construction in the area.

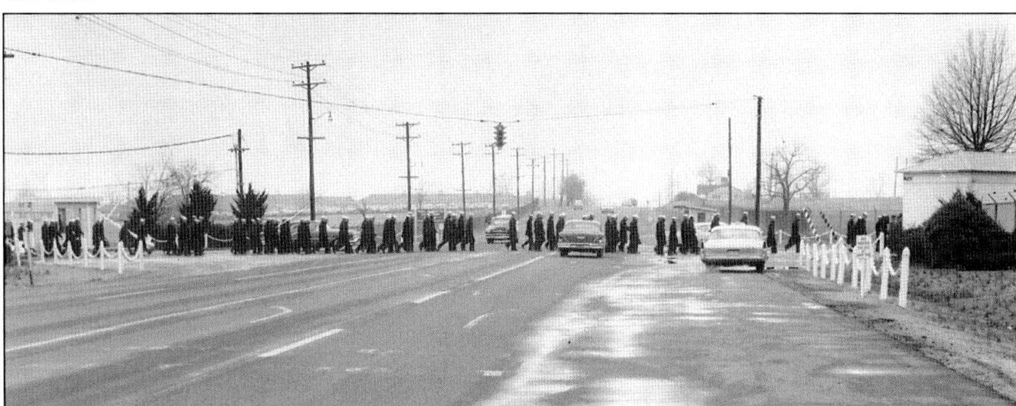

For more than two decades, traffic on Navy Road came to a standstill whenever sailors crossed from one side of the base to the other. In 1961, Millington mayor Tom Hall and county commissioner Rudolph Jones proposed a raised roadway and cloverleaf exit to the north and south gates. The reality was far simpler and cheaper: a raised walkway was constructed over Navy Road instead.

Two

Agriculture and Industry

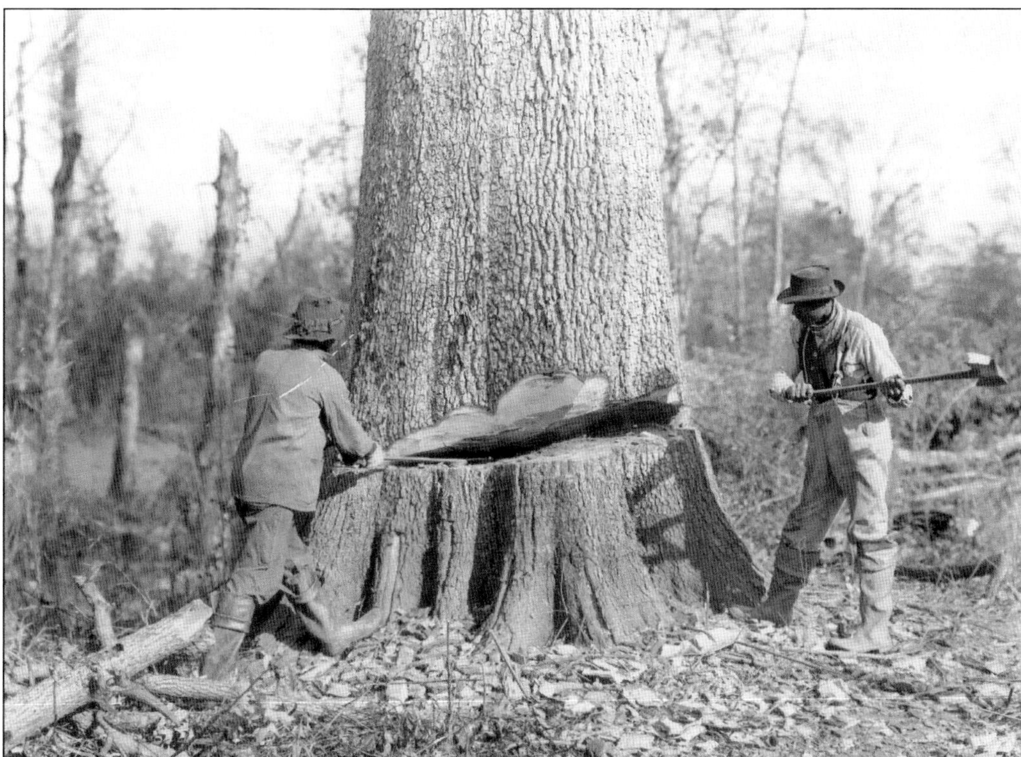

The first settlers of West Tennessee found vast expanses of virgin forest standing between them and building homes and farms. Some trees were so large, several axe-wielding loggers worked together to chop them down. Initially, felled trees were used for home construction, meaning that many early log cabins in the Big Creek area were quickly replaced with sawn lumber structures suitable to the wealthy planters that settled the area. (Courtesy MPL.)

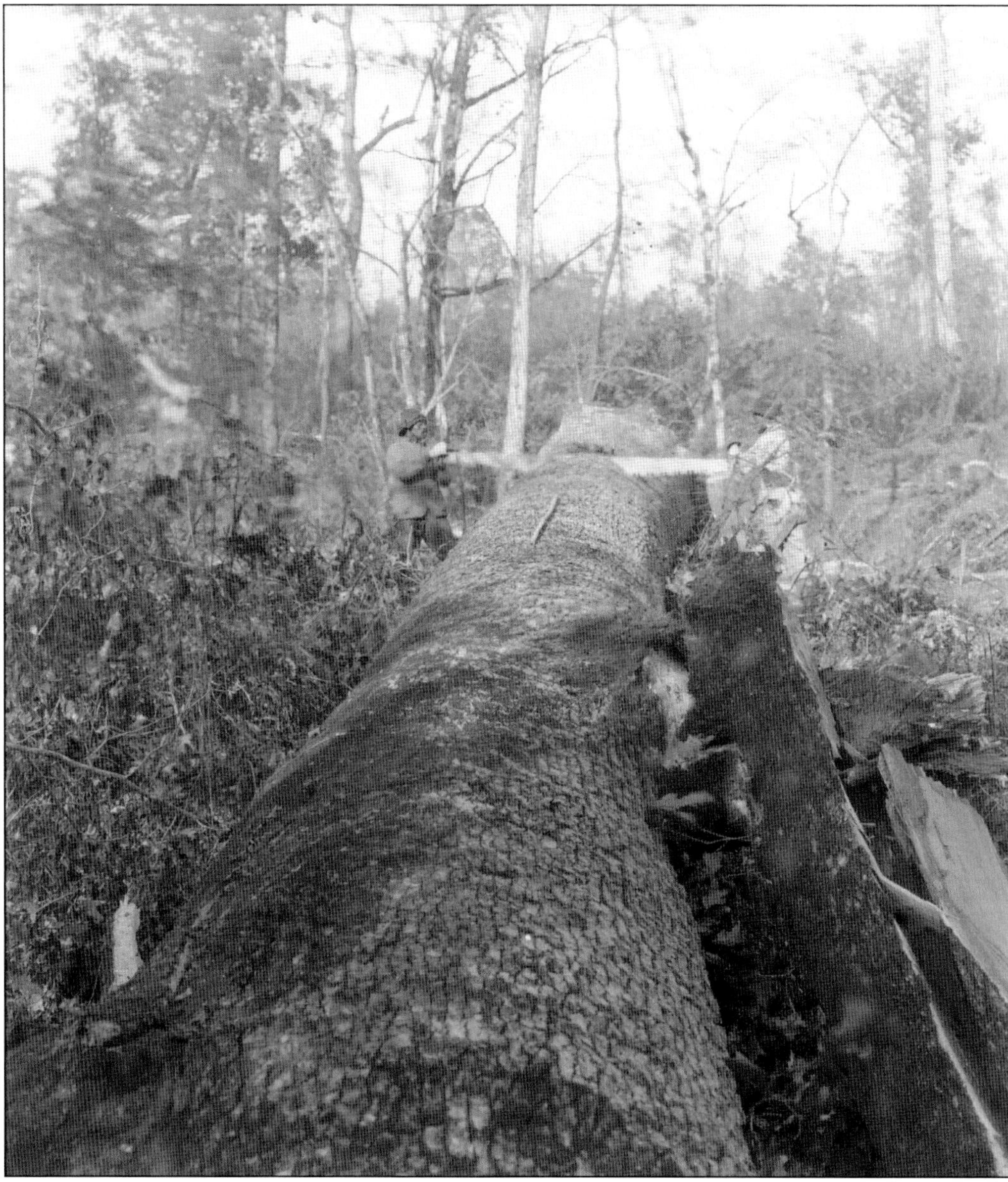

Sawyers operate a two-man saw in order to make this felled giant more manageable. After constructing their homes, landowners often donated trees for public buildings. For instance, William Sigler and William Ward made donations from their Big Creek properties to build the First United Methodist Church in Millington. Sigler's trees were also used to build a schoolhouse and Masonic Lodge in the Cuba area. (Courtesy MPL.)

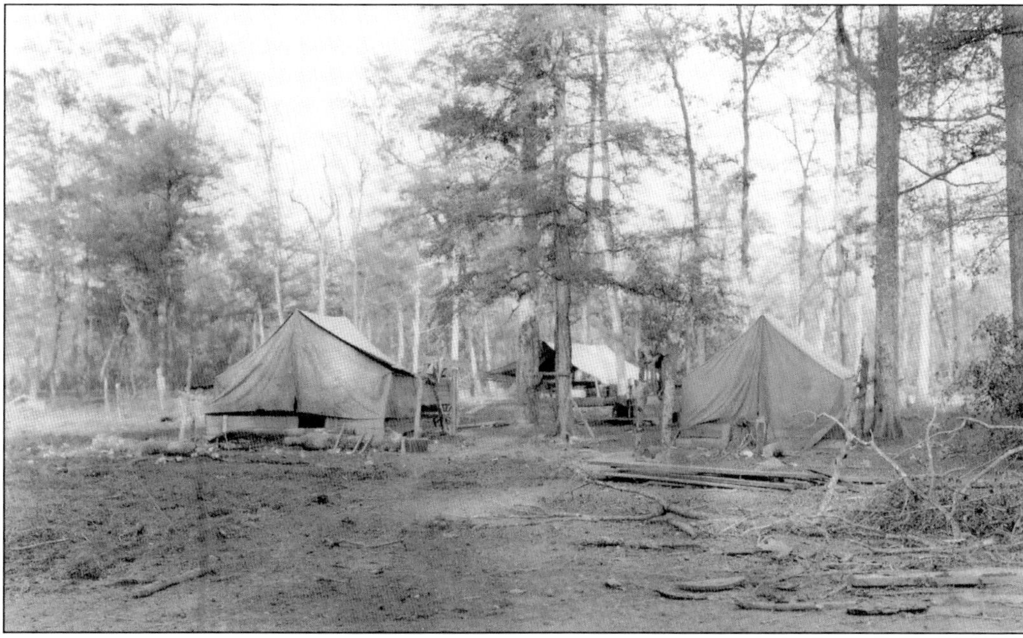

For more than a century, lumber camps like the one seen above dotted the landscape as loggers worked year-round to keep up with demand. The men lived in tents and cooked over open fires. However, beginning in the late 19th century, some areas established semipermanent logging camps with cabins for the crews. Occasionally, women and children accompanied loggers, who often lived a migratory existence with no permanent residence. African Americans provided the bulk of the workforce for many timber barons, as demonstrated below, both before and after the Civil War. *The Tennessee Encyclopedia of History and Culture* reveals that by 1912, lumbermen worked 60-hour weeks for as little as 14¢ per hour. (Both, courtesy MPL.)

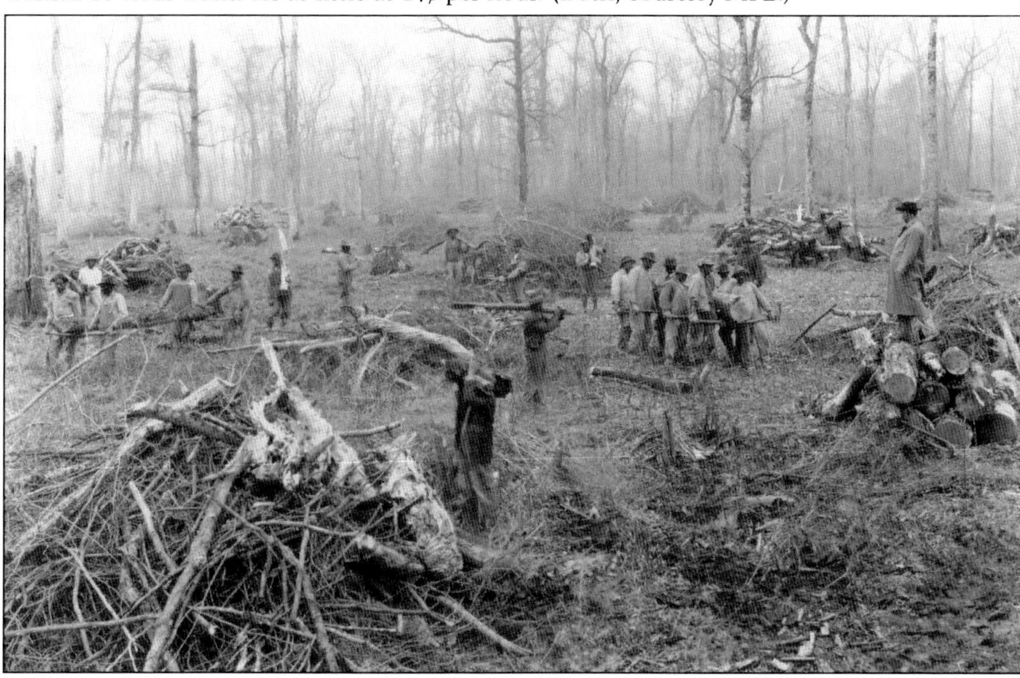

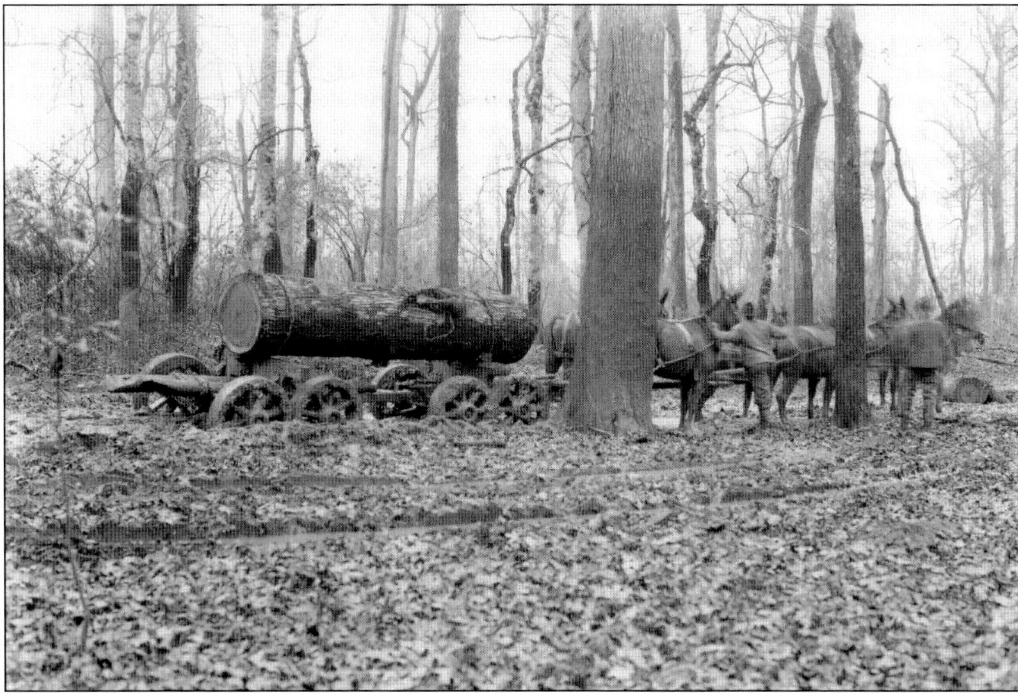

Enormous tree trunks were hauled, or skidded, through the often-soggy river bottoms by teams of horses, mules, or oxen. Deep ruts in the mud made such transport difficult. The term "skid row" originates with this industry, referring to the seedy businesses that catered to rowdy lumbermen. The advent of steam-powered cable skidders, shown below, somewhat simplified the job. High demand for lumber throughout the 19th and 20th centuries helped make Shelby County a leader in the hardwood industry but contributed greatly to problems of erosion in the Millington area. (Both, courtesy MPL.)

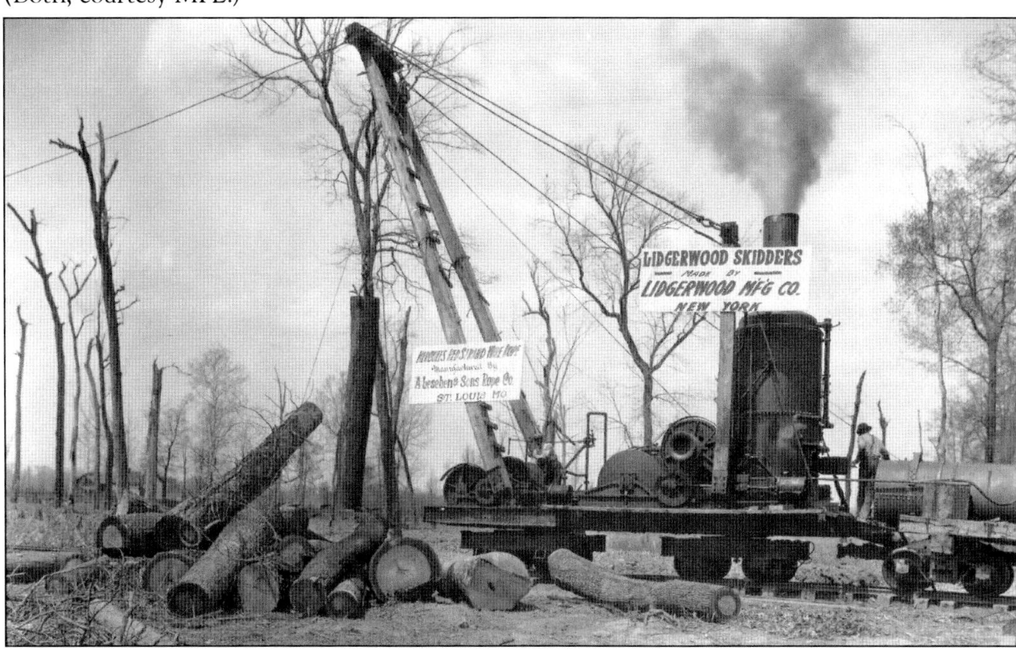

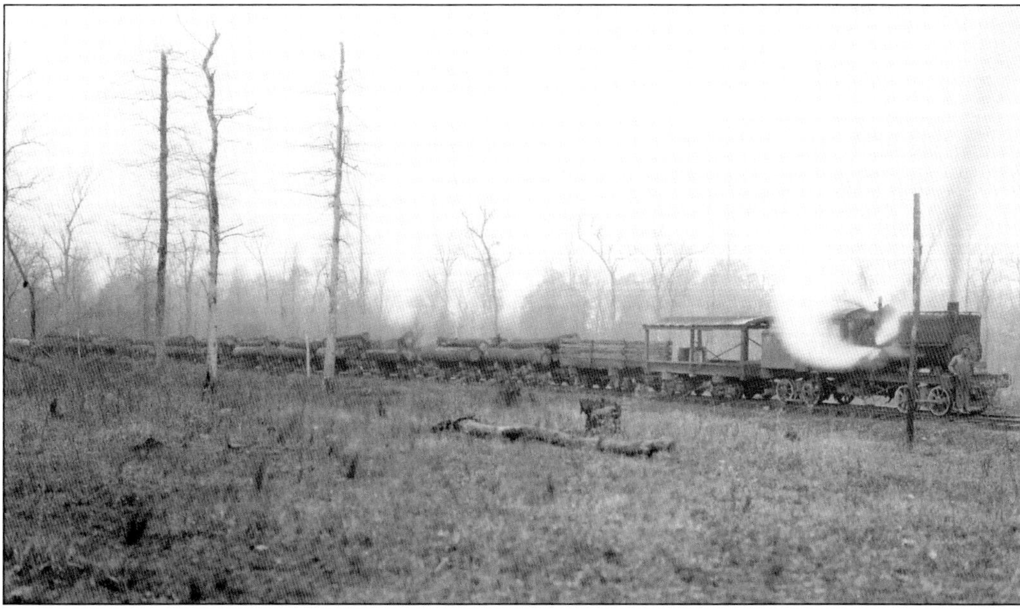

A steam locomotive laden with West Tennessee lumber passes through a once heavily wooded landscape, part of a journey so regular the cow standing only a few feet away from the tracks appears entirely unconcerned with the steam-spewing beast lumbering past. Stark, leafless branches of trees deemed unworthy by the lumbermen reach skyward as if attempting to escape the devastating deforestation. Another train, below, traverses the Illinois Central Railroad tracks that helped establish the town of Millington. A banner reveals the cargo and its destination, announcing "Red Gum" and "50 Cars for shipment to Europe." This shipment of "America's finest hardwood" was harvested by Lamb Fish Lumber Company of Charleston, Mississippi. (Both, courtesy MPL.)

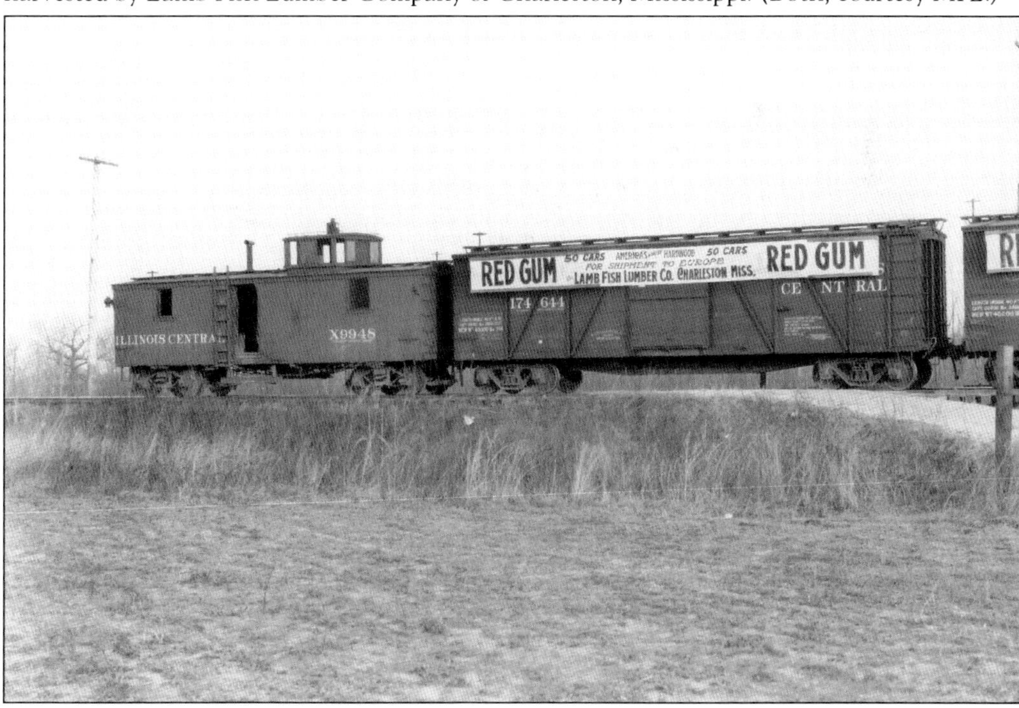

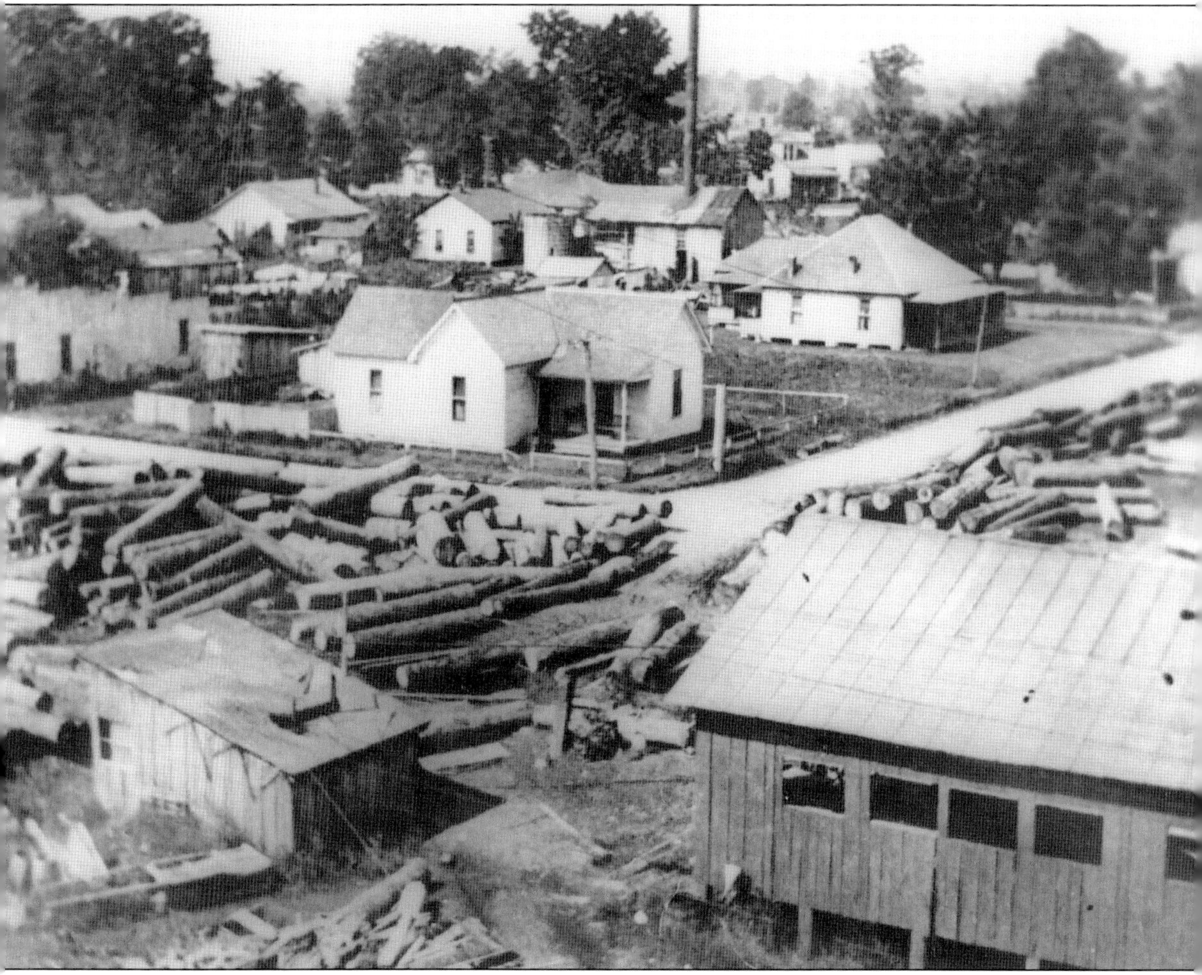

Eugene Woods was a boarder in the Tipton County home of farmer Will S. Hornsby in 1900. The 1969 sesquicentennial editions of the *Memphis Press-Scimitar* and the *Commercial Appeal* included a Lumbermen's Club of Memphis advertisement featuring Woods and his oxen team in 1892. His son Eugene Woods Jr. took over the business but left Millington before he was 35 to buy one of the most expensive homes on Stonewall Street in Memphis, although he maintained his interests in the sawmill. Hornsby never claimed to be anything but a cotton farmer, although he was both city judge and mayor of Millington for 12 years with the exception of one term. Hornsby's son Guy, however, told the 1920 census taker he was not only a timber man, but his occupation was "log buying." Hornsby and Woods Sawmill was conveniently located next to the railroad tracks in Millington. Trains bound for the sawmill could veer onto a dedicated sawmill spur between the depot and the E.A. Harrold store. (Courtesy MTC.)

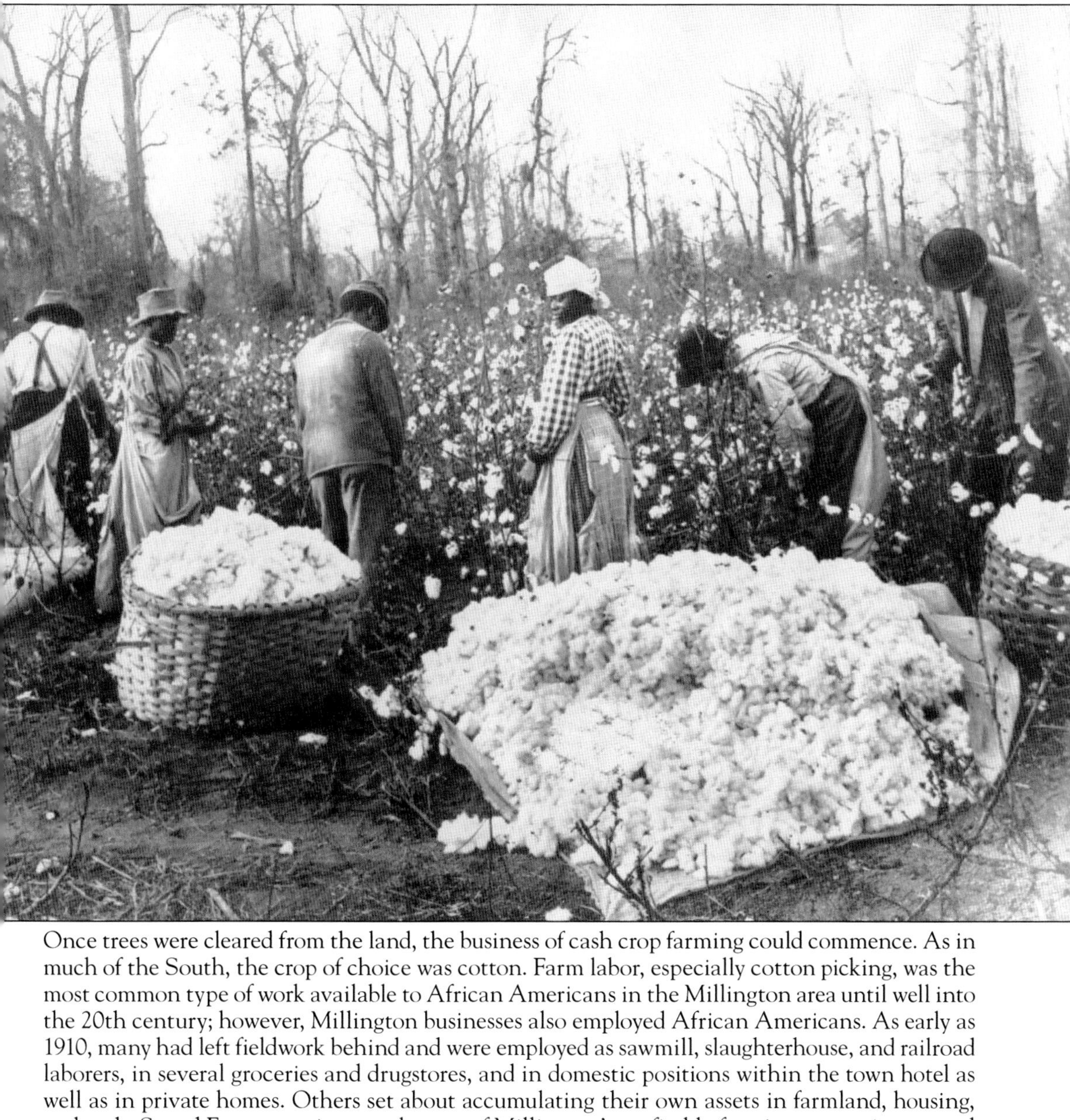

Once trees were cleared from the land, the business of cash crop farming could commence. As in much of the South, the crop of choice was cotton. Farm labor, especially cotton picking, was the most common type of work available to African Americans in the Millington area until well into the 20th century; however, Millington businesses also employed African Americans. As early as 1910, many had left fieldwork behind and were employed as sawmill, slaughterhouse, and railroad laborers, in several groceries and drugstores, and in domestic positions within the town hotel as well as in private homes. Others set about accumulating their own assets in farmland, housing, and cash. Sneed Farms continues to be one of Millington's profitable farming enterprises owned and operated by African Americans.

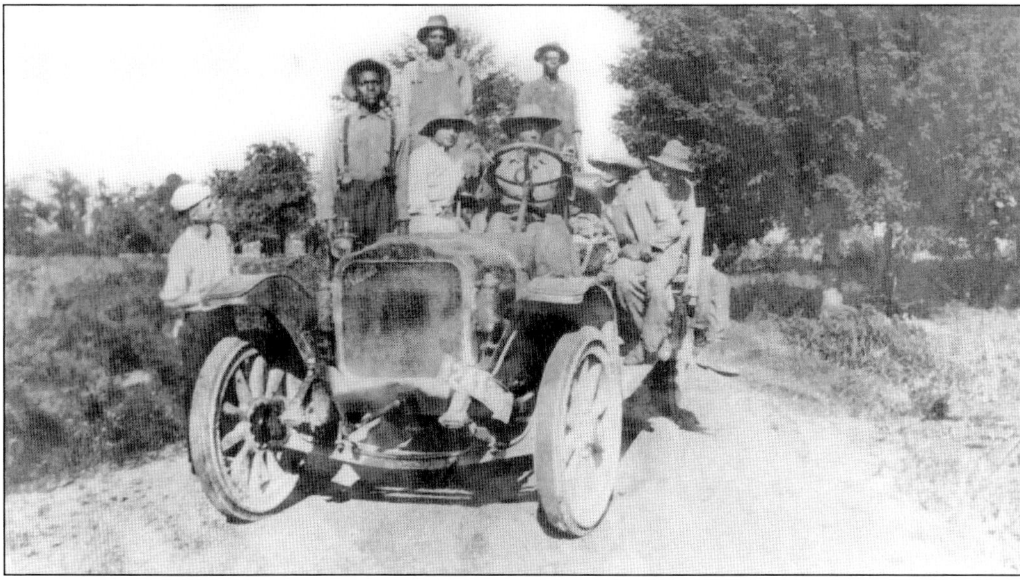

Before mechanized harvesting equipment lessened labor demands, day laborers ensured a timely harvest. Tenant farmers often hired their labor out to a farm owner, who would transport as many laborers as could find a seat or a spot to stand on a car's running boards. (Courtesy MTC.)

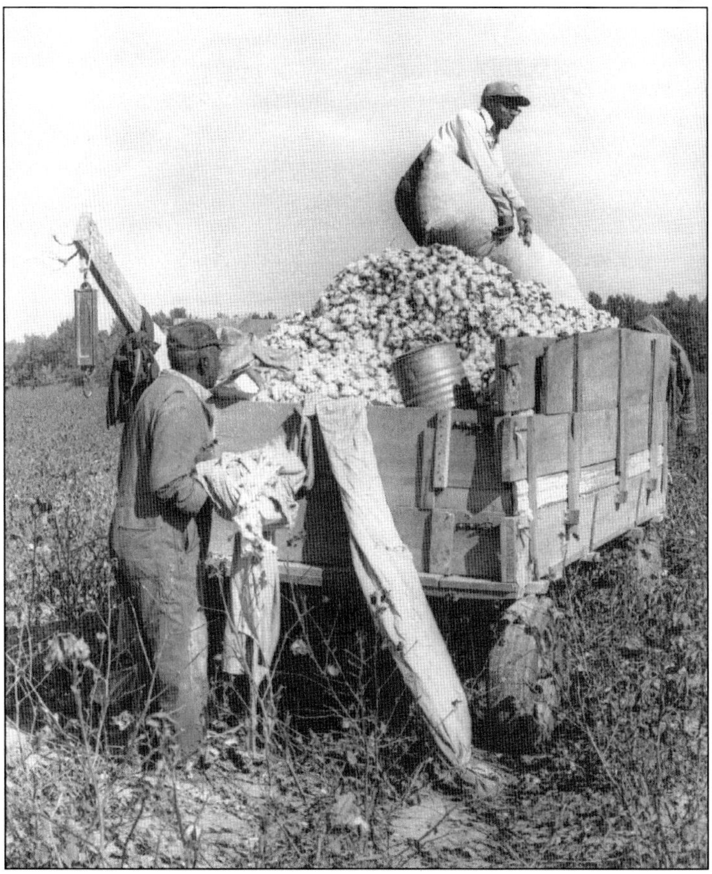

At harvest, cotton pickers filled long, narrow sacks like the one hanging from the back of the wagon. Before the sacks were emptied into the wagon, they were weighed on a scale, as seen hanging from a stick protruding from the back of the wagon. Knowing the weight of cotton delivered to the gin or compress protected the farmer from unscrupulous cotton agents that paid by the pound for cotton.

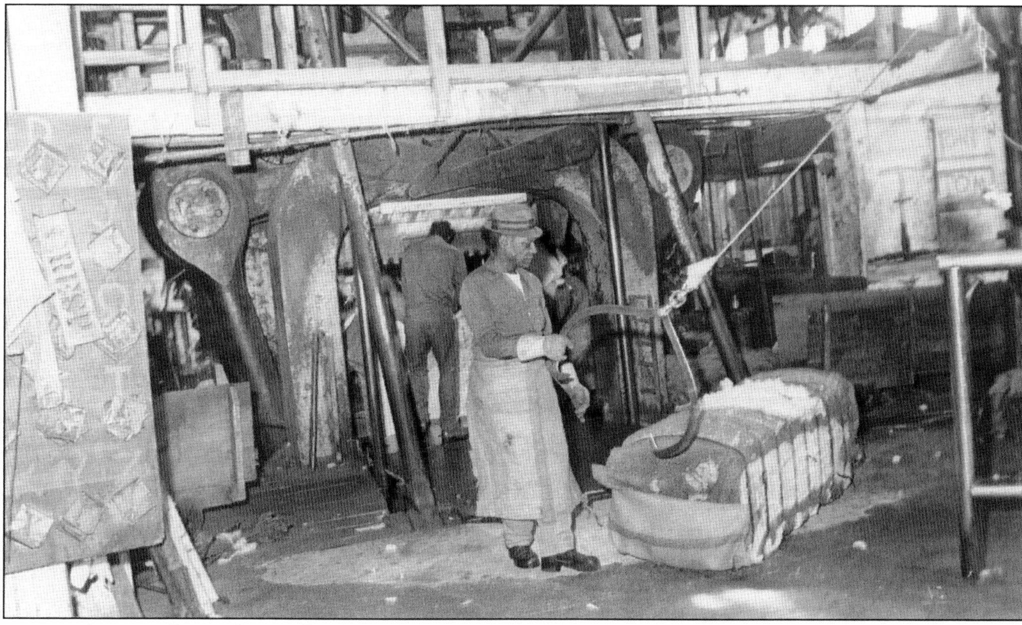

Many planters from the Millington area as well as from the countryside surrounding Memphis sent their cotton to Memphis Compress and Storage Company for processing into compressed bales, above, for warehousing or sale on the worldwide cotton market. The company, which was founded by Memphis millionaire William Neely Mallory and is now known as Mallory Alexander International Logistics, expanded its operations in the 1960s to handle diversifying needs beyond those of the cotton industry. Other planters stored their cotton at the Navy Yard Compress Company; this was not associated with the Millington naval base but was located where a Civil War–era Navy base stood in Memphis. Below, a farm laborer, photographed in 1964, hauls baled cotton to a dock perched on the roadside near a vast expanse of freshly harvested crops.

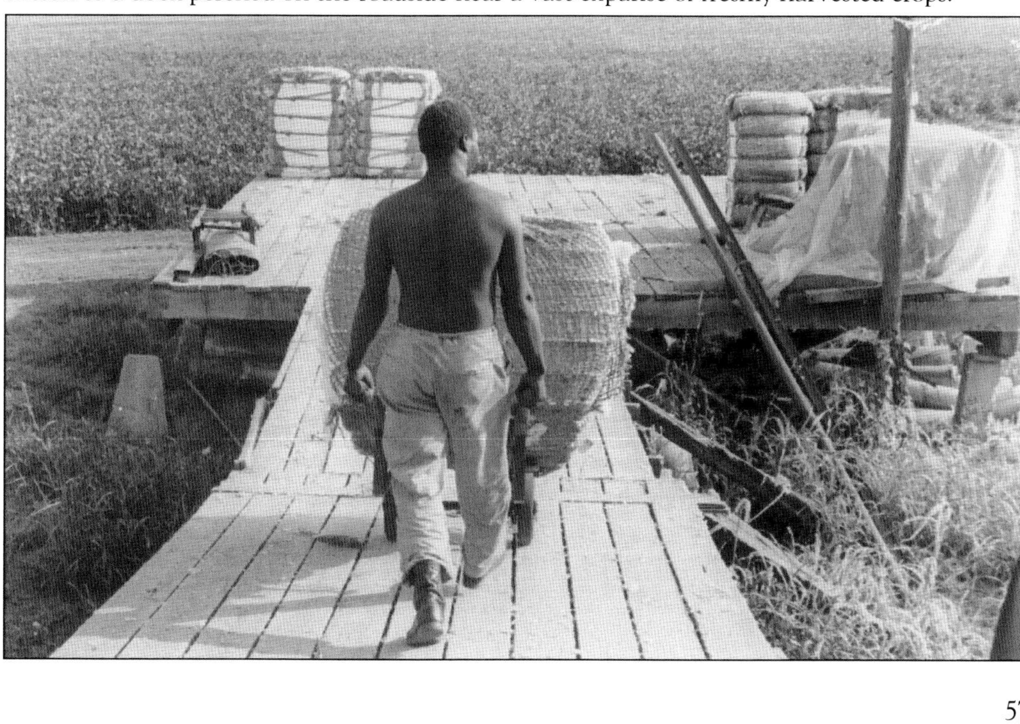

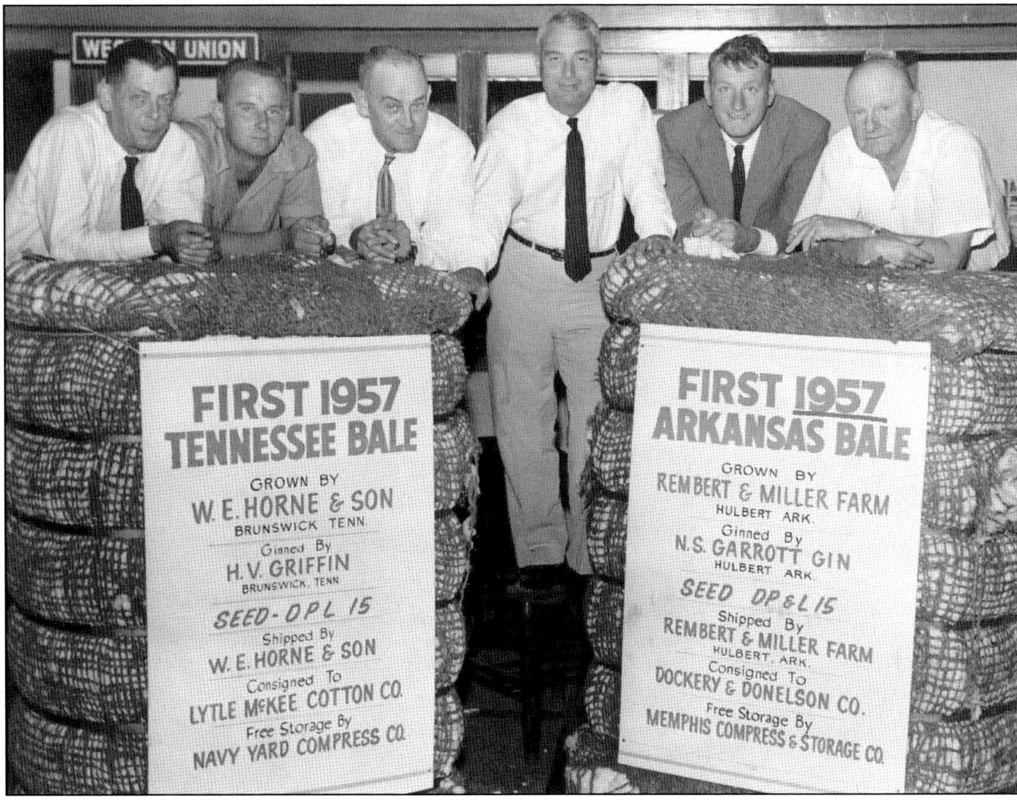

As with any business, cotton prices decline with supply and increase with demand, resulting in an annual race to the Cotton Exchange. The local winners in 1957, pictured above, were W.A. Horne (far left) and his father, W. E. Horne (third from left), who sold that year's first bale at 75¢ a pound. Sam Rembert Jr., standing between the bales, and his partner Dave Miller (far right) grew the first bale sold out of Arkansas. By the time this picture was taken in late August, prices were down to 30¢ a pound. By 1971, W.A. "Buddy" Horne and his brother Billy were in charge of Horne Brothers Farms. Below, Johnny Kilpatrick idled his employers' John Deere tractor while Buddy Horne (right, pointing) discussed the year's crops with E.S. Permenter (center).

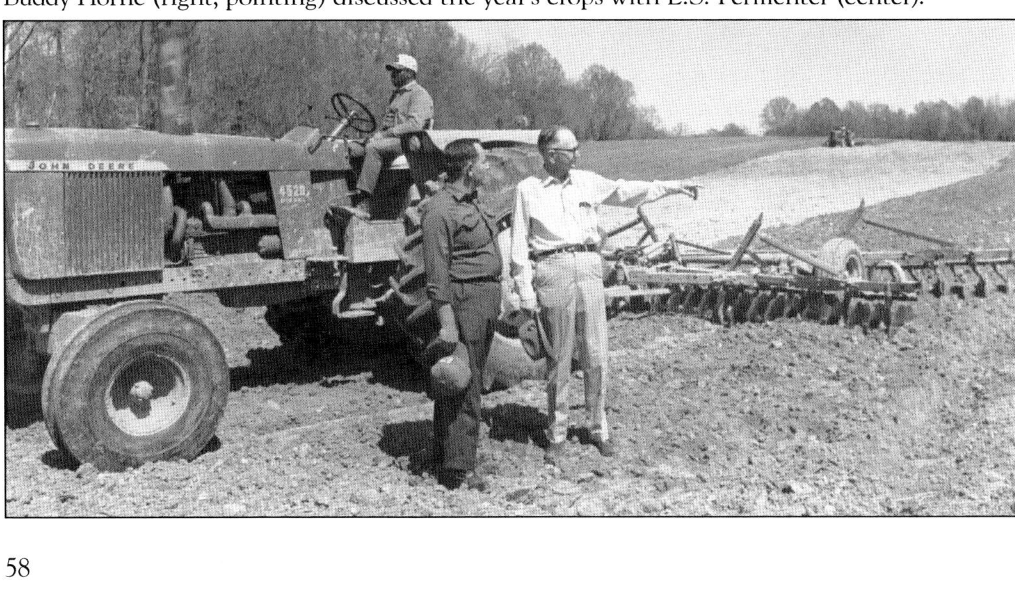

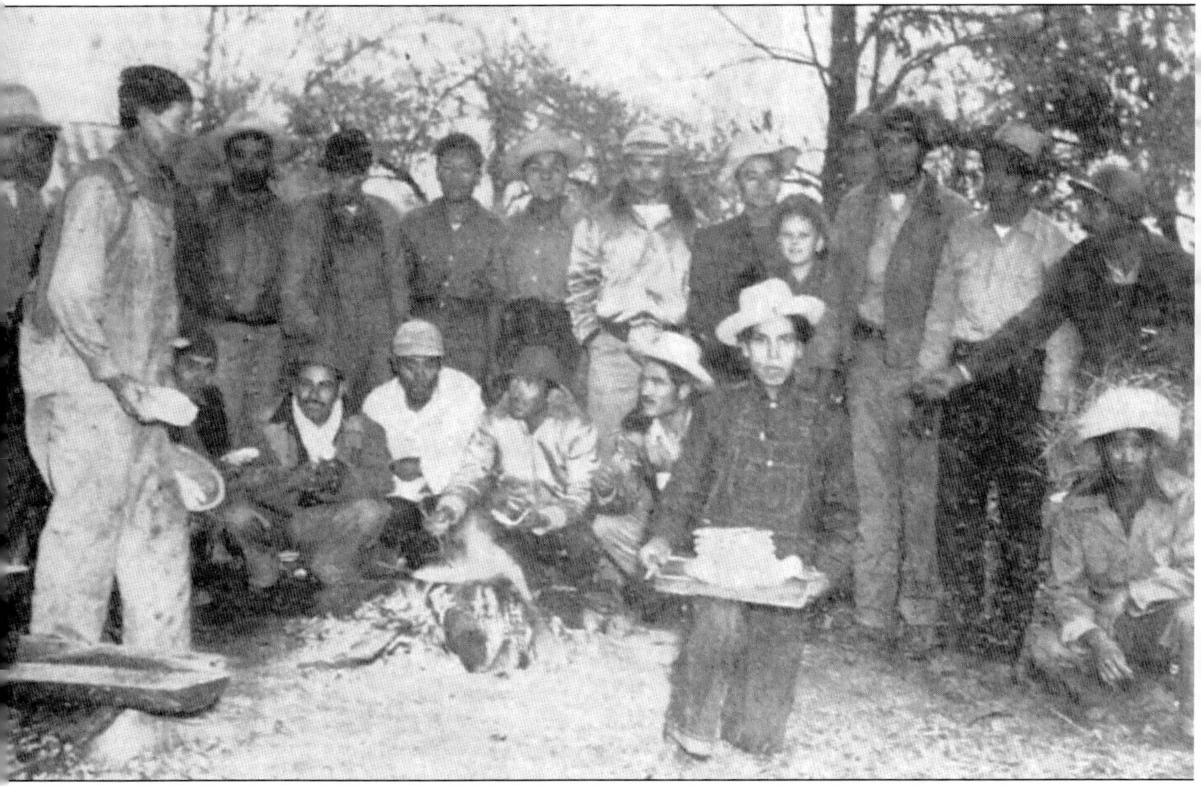

As late as 1940, when the powder plant land purchases occurred, E.A. Harrold had 100 tenant families living on his land west of town. After the powder plant displaced tenant farmers and landowners alike, yet brought thousands of jobs to the area, new sources of farm labor were needed. Millington farmer P.A. McNamara brought the first Mexican farm laborers to Shelby County in 1951. While Arkansas had some 23,000 and Mississippi had about 3,000 Mexican laborers, Tennessee had only a few hundred—and 25 of those were in Millington.

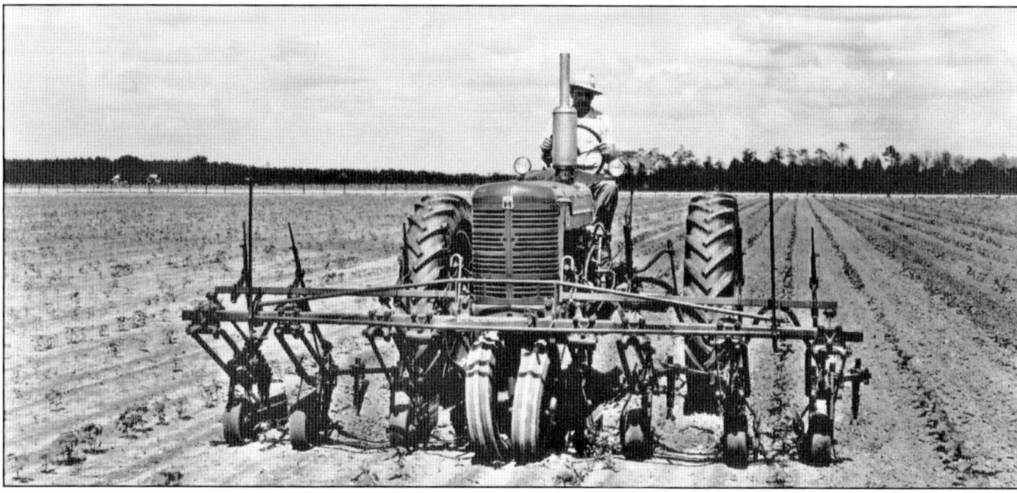

By the 1960s, much of the work of cultivating young cotton plants was accomplished by tractors, seen above, and other mechanical equipment rather than by hand. Forced to seek out other means of earning a living, many farm laborers moved to Memphis to seek jobs in the city. As a result, farmers such as Cliff Delashmit had trouble finding laborers to work the fields. Delashmit even offered the house seen below rent-free for use by a willing tenant farmer, but it was to no avail.

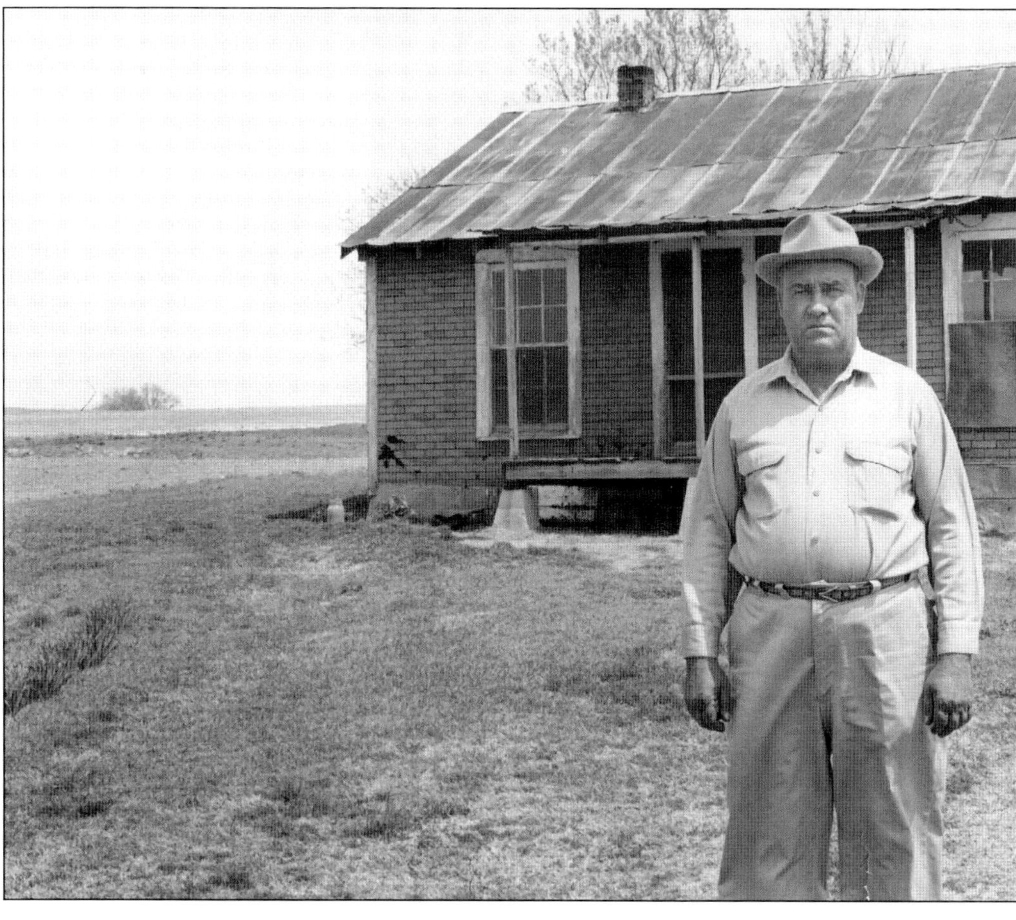

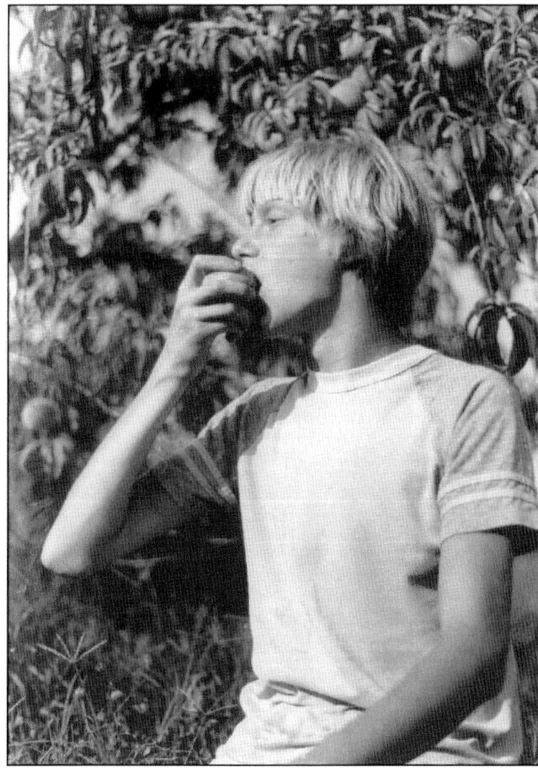

There is little as delicious as fruit eaten within moments of picking, as this young man has learned. For three generations, Jones Orchard has offered that experience to its customers, who can pick their own or purchase harvested fruits from roadside stands flanking the orchard or from farmer's markets in Memphis. The orchard produces several varieties of peaches, nectarines, plums, strawberries, blackberries, blueberries, watermelons, pumpkins, and cantaloupes, along with vegetables, such as sweet corn and Vidalia onions. After convenience foods replaced the practice of canning bulk produce, Jones Orchard began to offer their own bottled jams, jellies, and dressings. In recent years, Jones Orchard has added a restaurant, homemade fruit breads, and a Halloween favorite, the corn maze, to its repertoire.

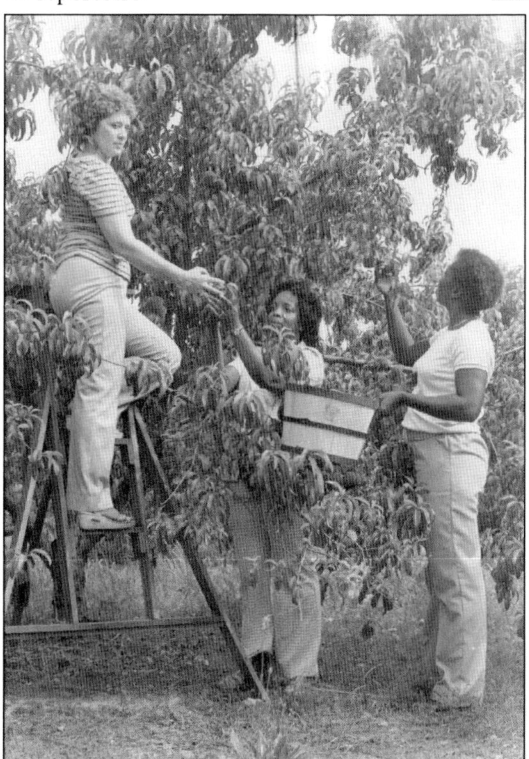

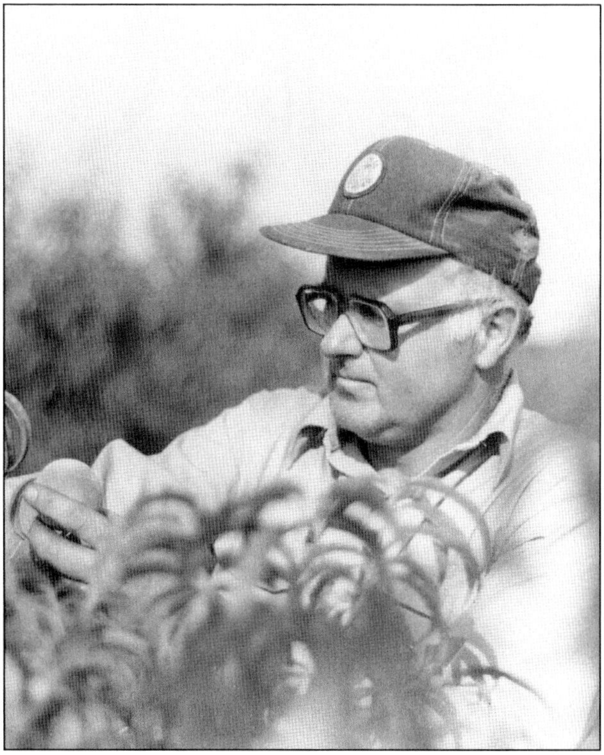

Lee Wood Jones inherited a thriving orchard from his father, H.L. "Peaches" Jones, who got his start with mule-plowed fields and door-to-door fruit sales in the 1940s. After earning a degree in agriculture from the University of Tennessee at Martin, Lee Jones expanded the 600-acre farming operation by using tractors and other modern techniques to grow many varieties of fruits and vegetables. He has also grown wheat and soybeans but is best known for his pick-your-own fruit orchard. Below, Jones and his wife, Juanita, began teaching their children the family business early on. Their son Henry grew up to contribute innovative marketing techniques to Jones Orchard, carrying a family tradition into its third generation.

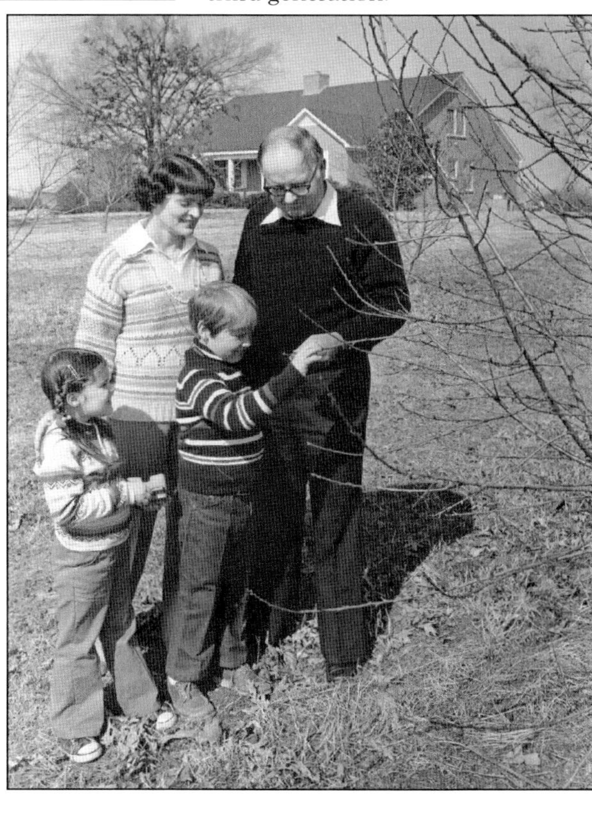

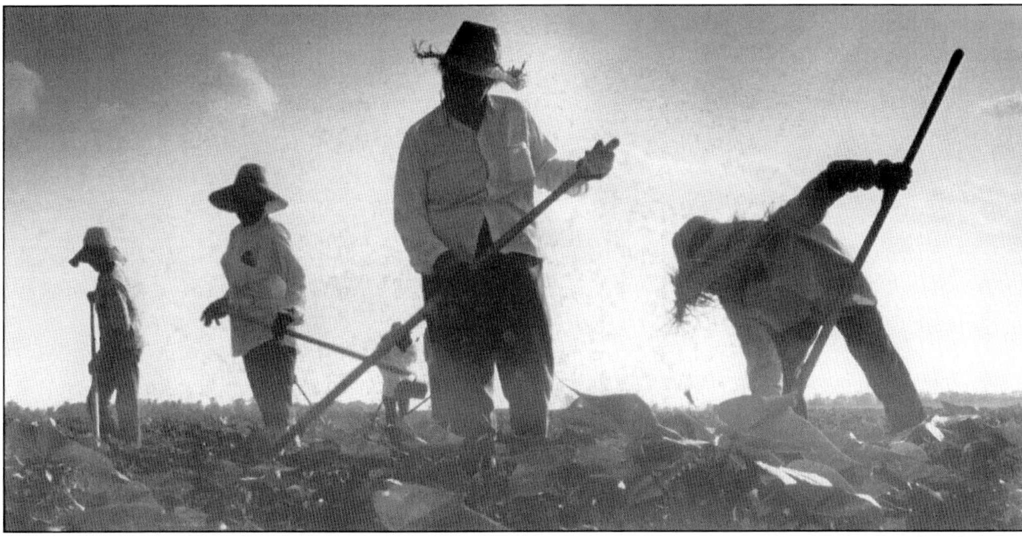

Scenes like this captivating 1965 photograph, taken as storm clouds threaten the landscape, have been a common sight throughout the area for generations. Large farm families often worked together in the fields to produce their food as well as cash crops. The Sneeds are third-generation owners of Sneed Farms, which received the 2009 Shelby County Conservation Farmer of the Year Award for best soil conservation practices and was featured in the 2010 America's Farmers advertising campaign. Over the years, Sneed Farms has raised cotton, wheat, corn, soybeans, sorghum, sunflowers, and livestock on some 9,000 acres. Below, Betty Clark earned the nickname "Lima Bean Queen" after she married Herbert Sigler, a fifth-generation farmer. On their Quito Road farm, they raised beef cattle amidst 1,200 acres of soybeans, cotton, wheat, and of course, lima beans, which they gave away freely to friends and neighbors.

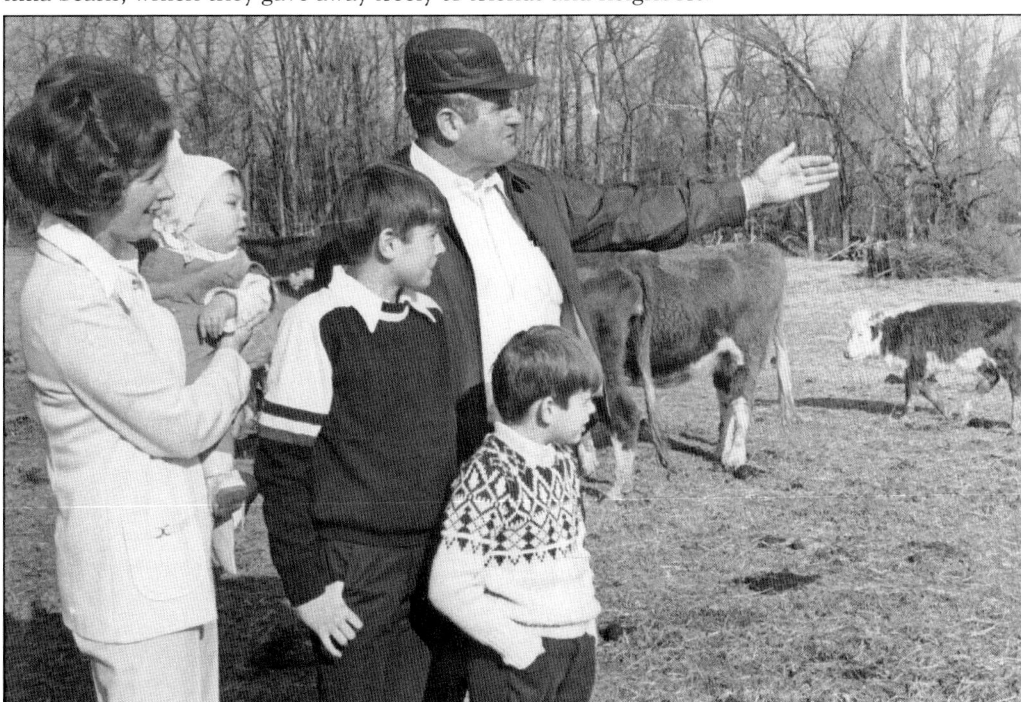

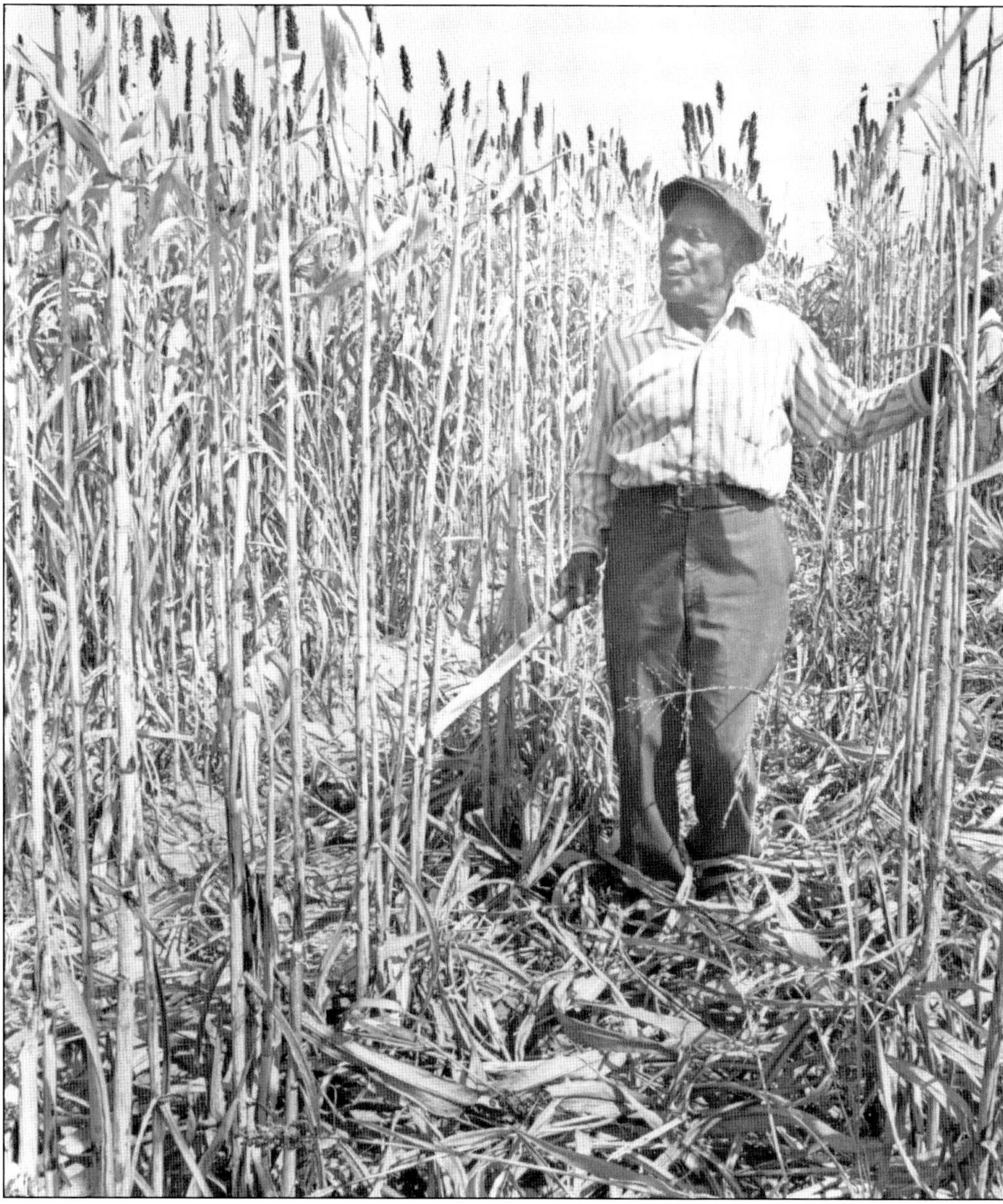

Some historians believe that the first domesticated sorghum in the South was an unintended by-product of slave importation, as the cane is indigenous to Africa. In the 1850s, an influx of cane and seeds imported from China, France, and South Africa resulted in the distribution of thousands of federally patented sorghum seeds to American farmers. Although sorghum is drought-tolerant, early wild varieties thrived in flood-prone areas like Big Creek. Here, a sorghum farmer, whistling while he works, assesses his stalks to determine which are ready for cooking. An important first step in the process is stripping the leaves, which he does with a machete, leaving a crunchy carpet of shorn leaves amidst the stalks. Next, the stalks will be cut and the seed heads removed. Finally, the stalks will be loaded onto a truck or wagon and transported to the mill for grinding, boiling, and skimming before the sweet nectar is ready for eating or for sale. (Courtesy *Millington Star.*)

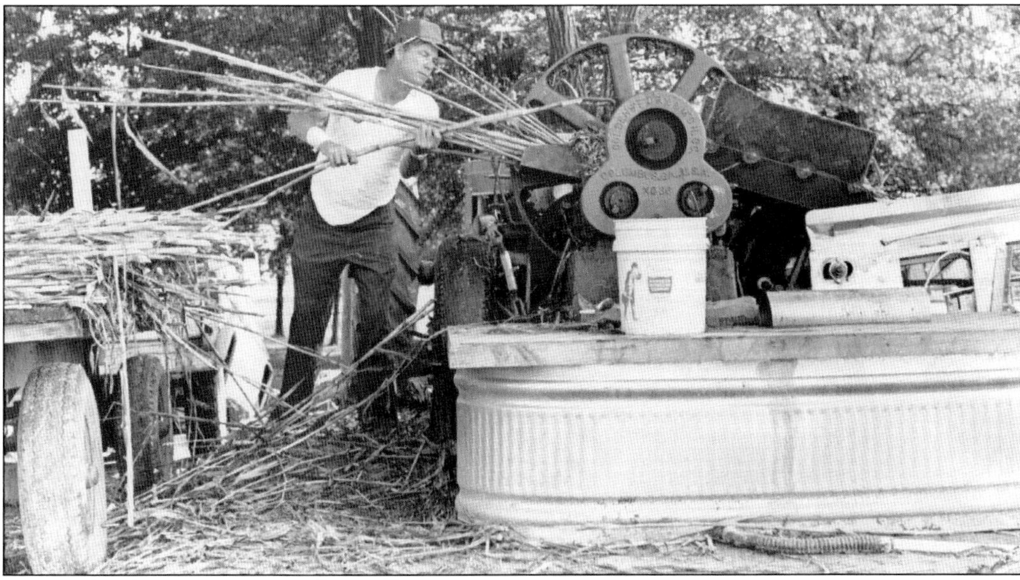

W.S. "Babe" Howard, owner of Lach-Host Farms, instructed his employees to always feed the sorghum canes into the mill the hard way—big end first—to avoid a breakdown. Above, Albert O'Neal feeds canes into the c. 1918 mill Howard bought from a farmer in Oakland, Mississippi. Below, Erskine Maples tends to Myrtle the mule, who was threatened with retirement when the health department informed Howard that mules were not allowed to turn the grist because they drew flies. Howard officially replaced Myrtle with a tractor, but for nostalgia's sake, he kept the mule working for all to see as they passed the sorghum operation on their way to the Wal-Mart across Highway 51.

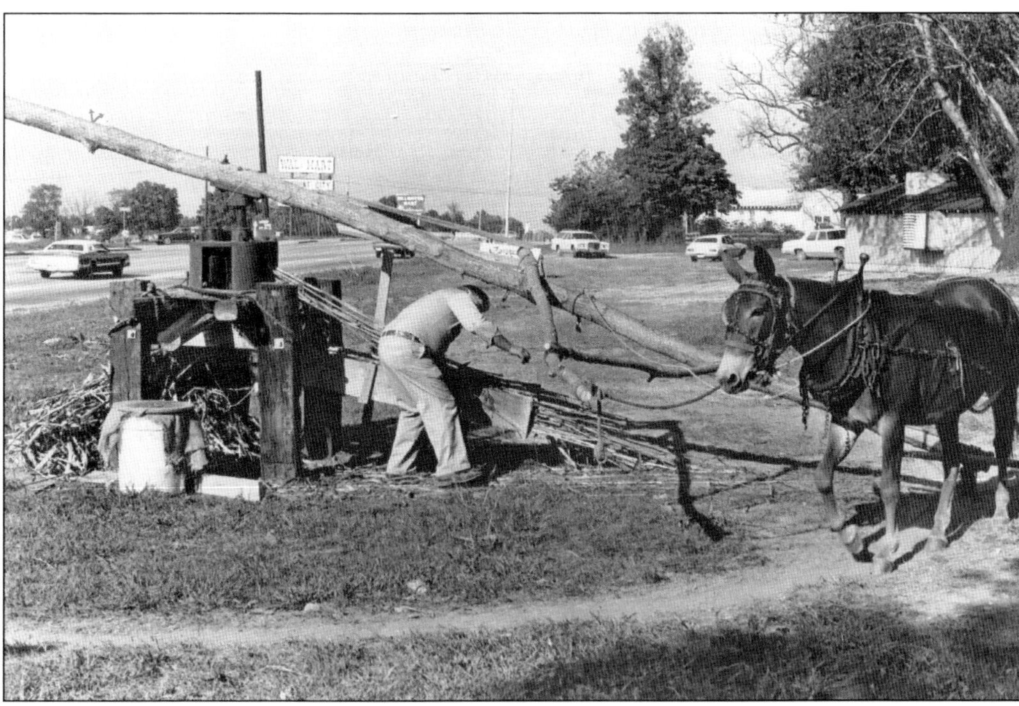

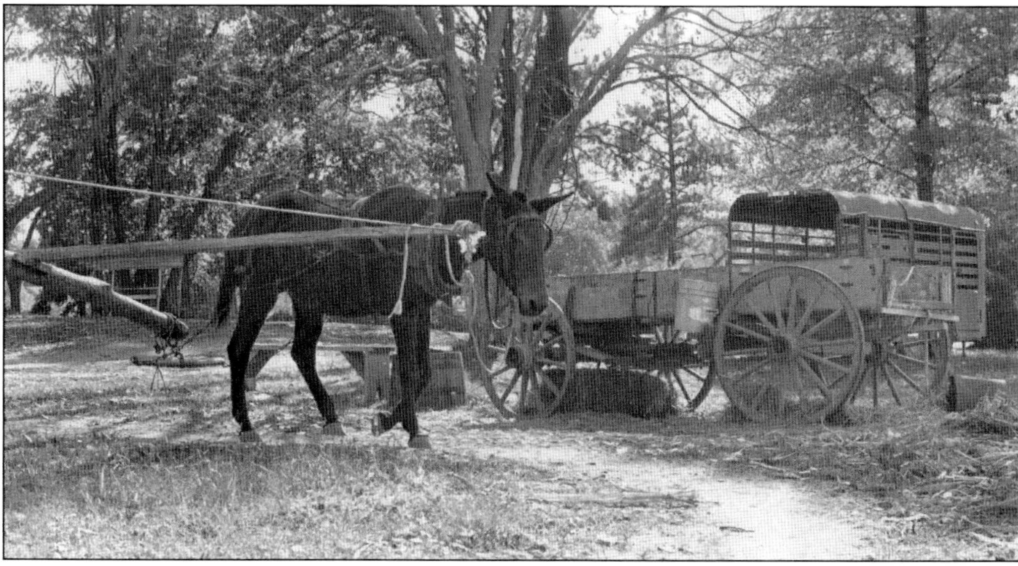

Myrtle the mule, above, patiently plodded her well-worn circle around Babe Howard's sorghum mill. Meanwhile, Stuart Howard and his sister Charlotte Thompson, pictured below, had the unpleasant task of skimming the froth off the cane juice as it boiled in the custom-built stainless steel cooking pan. Sorghum cane juice begins to sour about three hours after it is pressed, so cooking and cooling must proceed quickly. Molasses quality is judged in part by its color, which deepens with length of cooking, and Lach-Host Farms was known for its pure, honey-colored molasses. The farm, named for Howard's children (Laura, Charlotte, Holly, and Stuart), went out of business after 35 years following Howard's death in 2008.

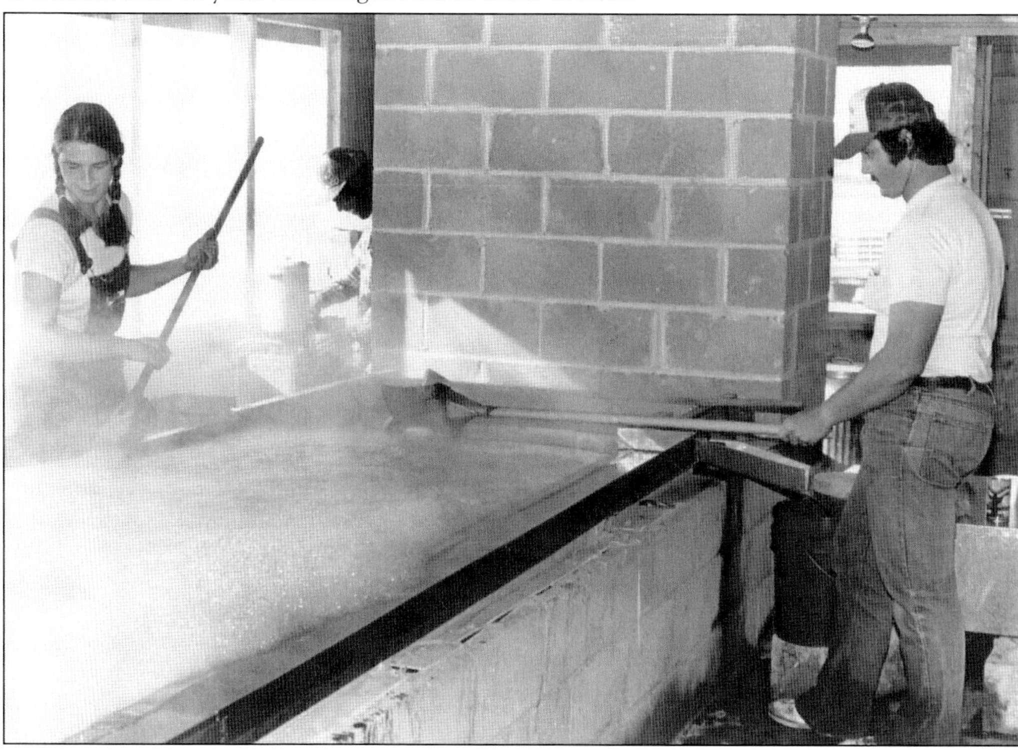

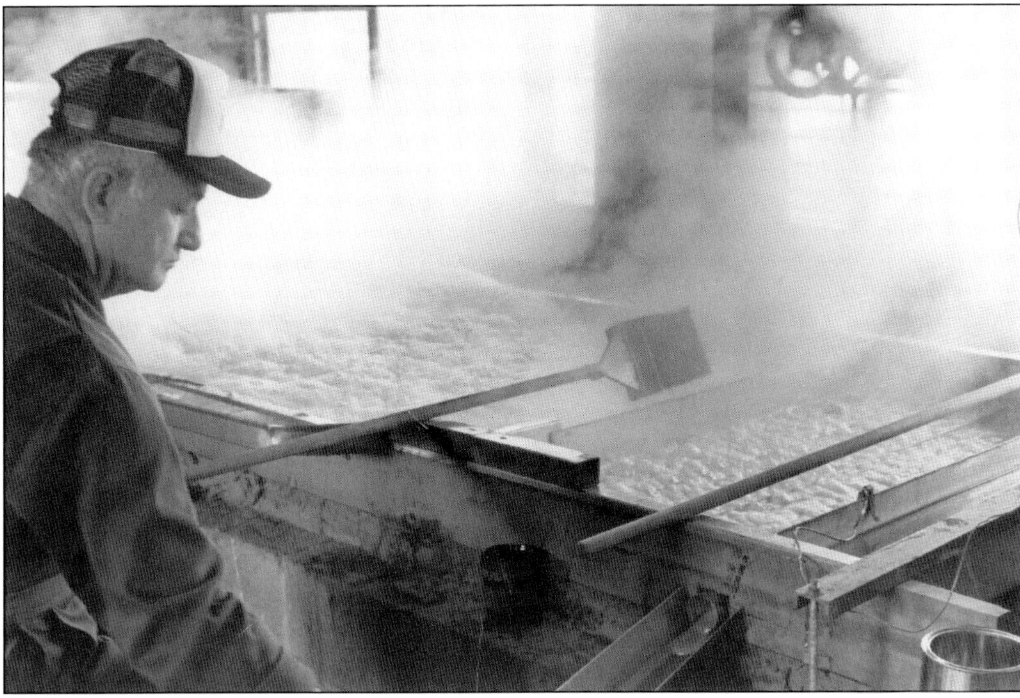

Finally, the fruits of an arduous process are realized as a well-trained sorghum cooker prepares to fill a container with golden molasses. Before World War II, many cookers drained their product into half-gallon or gallon cans similar to the paint can in the bottom right corner of this photograph. Molasses making is a labor-intensive process, and today, most molasses is sold by the quart. (Courtesy *Millington Star.*)

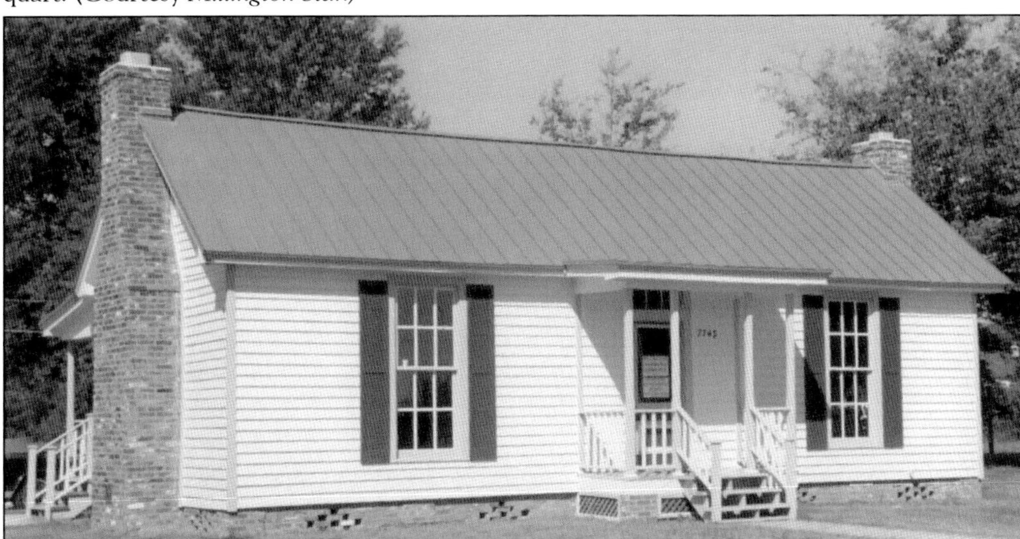

The Jack Easley house, once home to Val Williams Insurance Agency, was the town's oldest residential building when Babe Howard rescued it from destruction. Howard, the first president of the Millington Chamber of Commerce, moved the building from its original location on Easley Street to its current site on Church Street. Pictured here after renovations, the structure now houses the Millington Chamber of Commerce offices. (Courtesy MTC.)

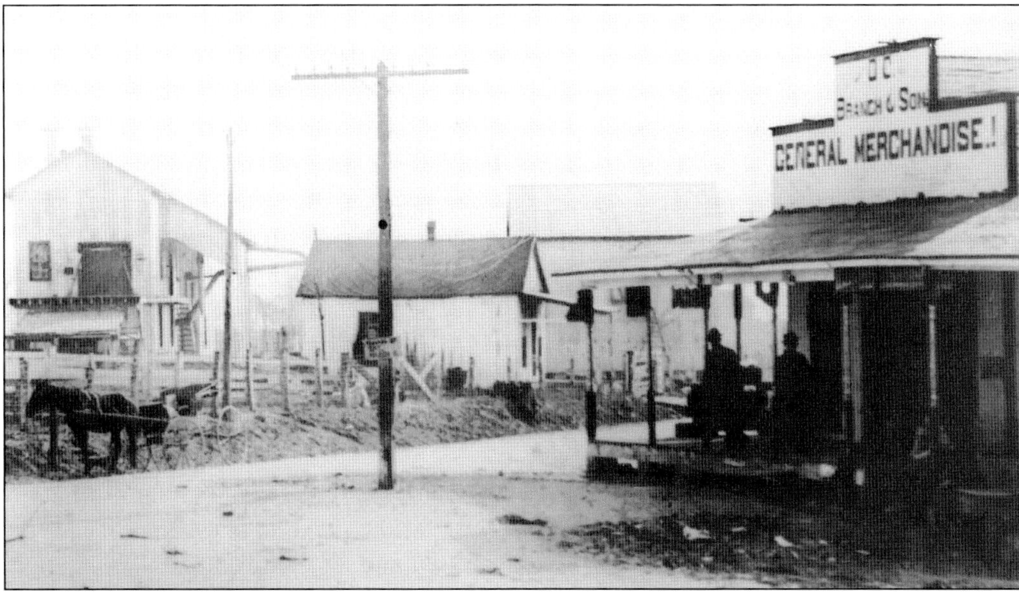

Oliver C. Branch was one of Millington's early businessmen. He operated the general store pictured here on the right and the cotton gin on the left. There were two other cotton gins, owned by W.A. "Buck" Sigler and E.A. Harrold. From 1905 to 1924, Sigler also owned a store in a building shared with the Bank of Millington. The two-story building had an elevator to take him and his family upstairs where they lived. Downstairs, groceries were sold from one side of the store and dry goods, as seen below, were sold from the other side. Pictured here around 1910 from left to right are Sigler, Arthur Smith, and John Ed Wilson. (Both, courtesy MTC.)

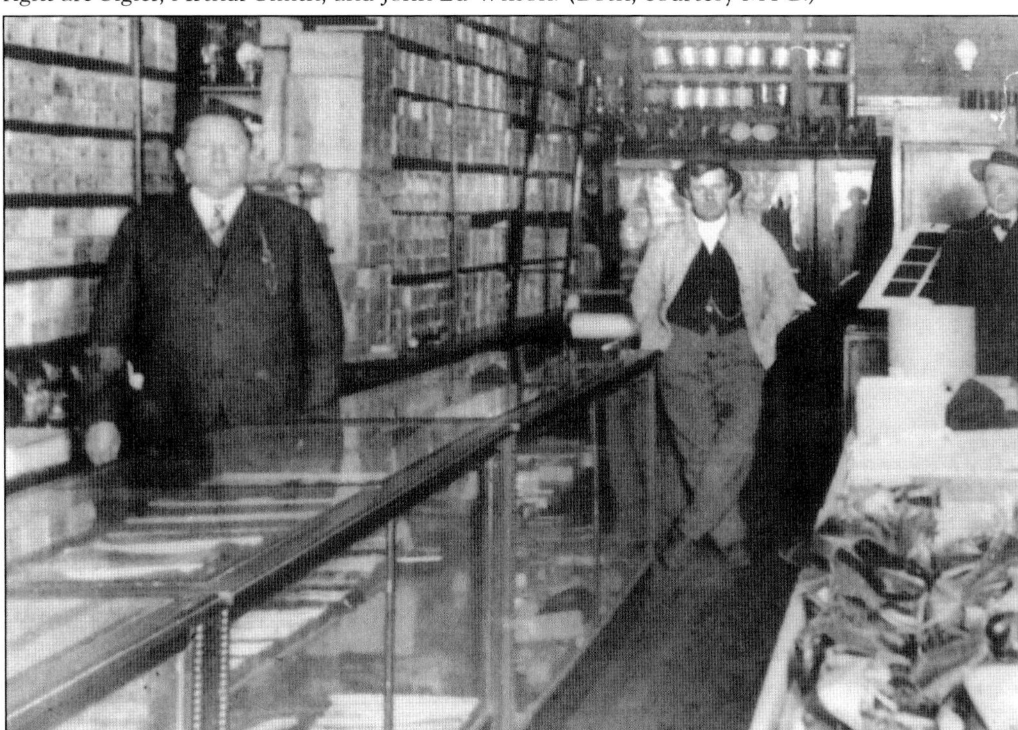

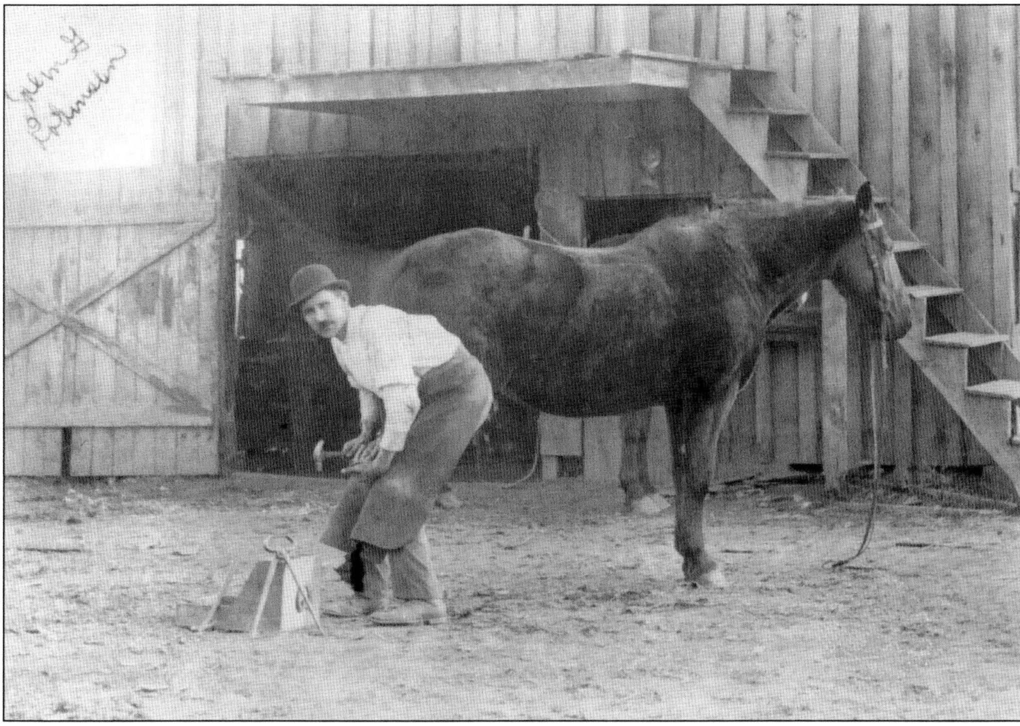

John G. Robinson, pictured above, was one of Millington's three blacksmiths in 1910. His son John T. Robinson, along with Jack Younger, manned the blacksmith shop below. Younger's father died before the boy was seven years old, and he and his mother lived with his grandparents, Ed and Mary Black. After apprenticing a while under Robinson, Younger went to work in a Memphis iron shop and later joined the Army. (Both, courtesy MTC.)

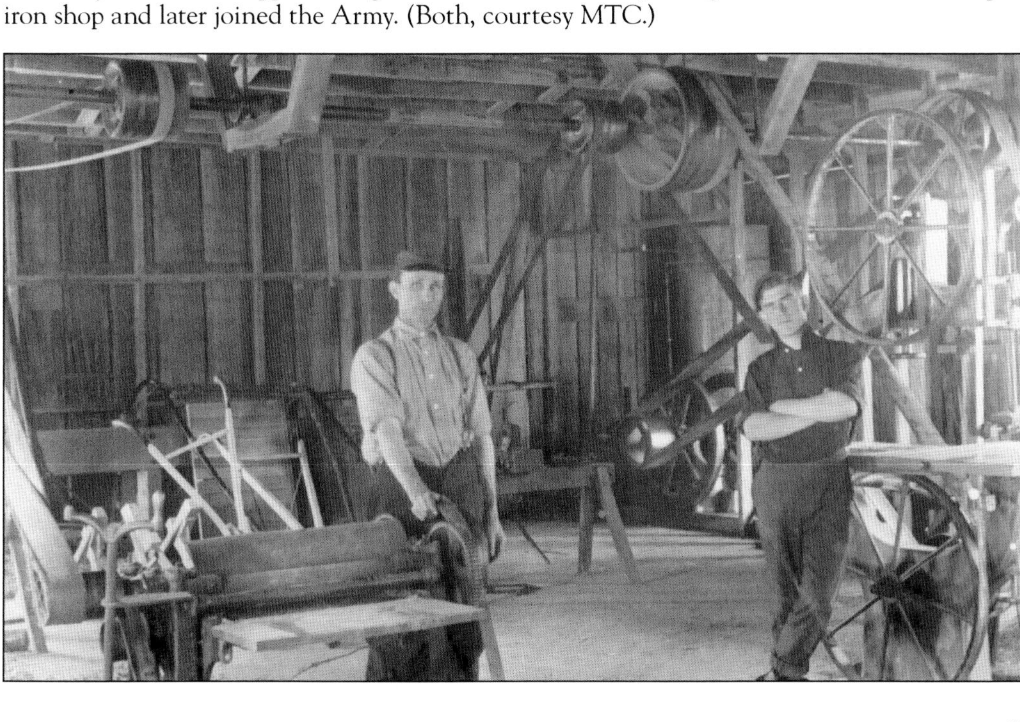

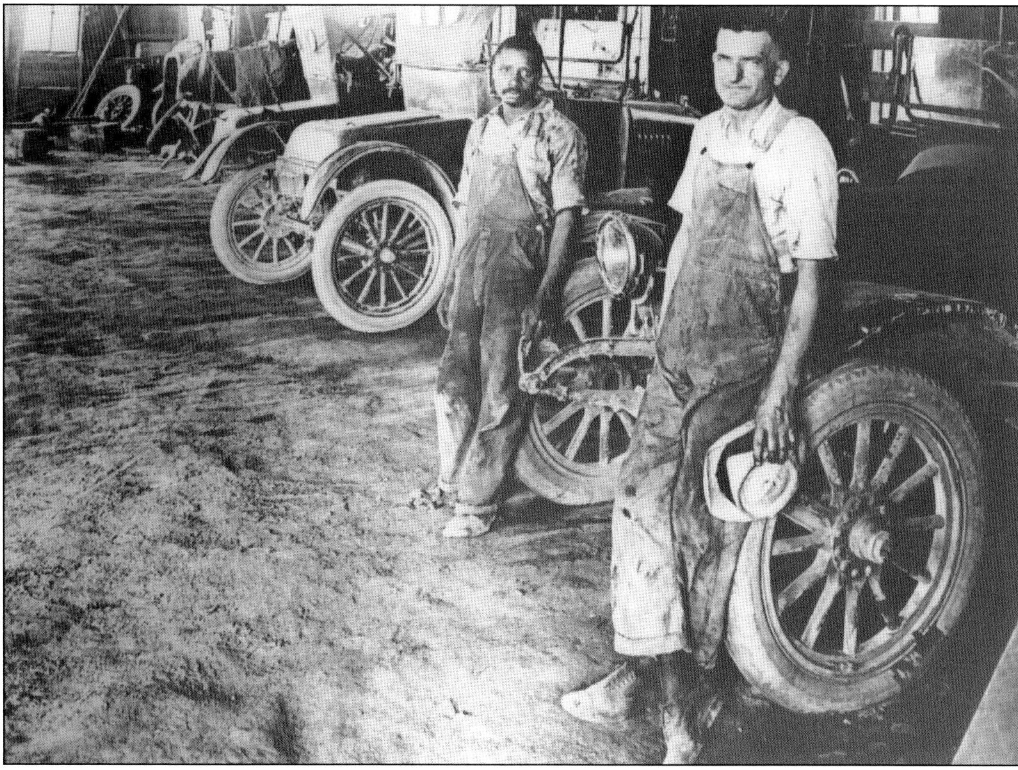

Frank and Charlie Hill were two of the town's first automobile mechanics. Hill's Garage was established about 1916 on Easley Street. Pictured here in 1919 are Frank Hill and employee Ike Hillary (left), who was killed the next year while working for the power company. By 1930, the garage was home to Hugh Thompson's Ford dealership. Below, a young Babe Howard visits with John L. Robinson, who worked at the Ford garage. Robinson's grandfather was the blacksmith John G. Robinson. (Both, courtesy MTC.)

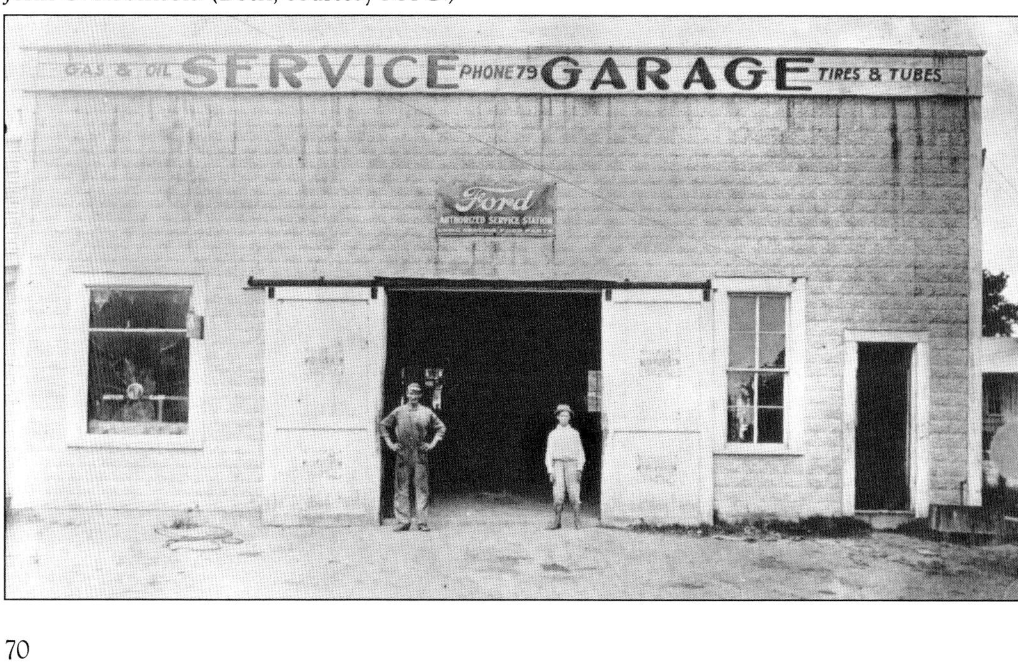

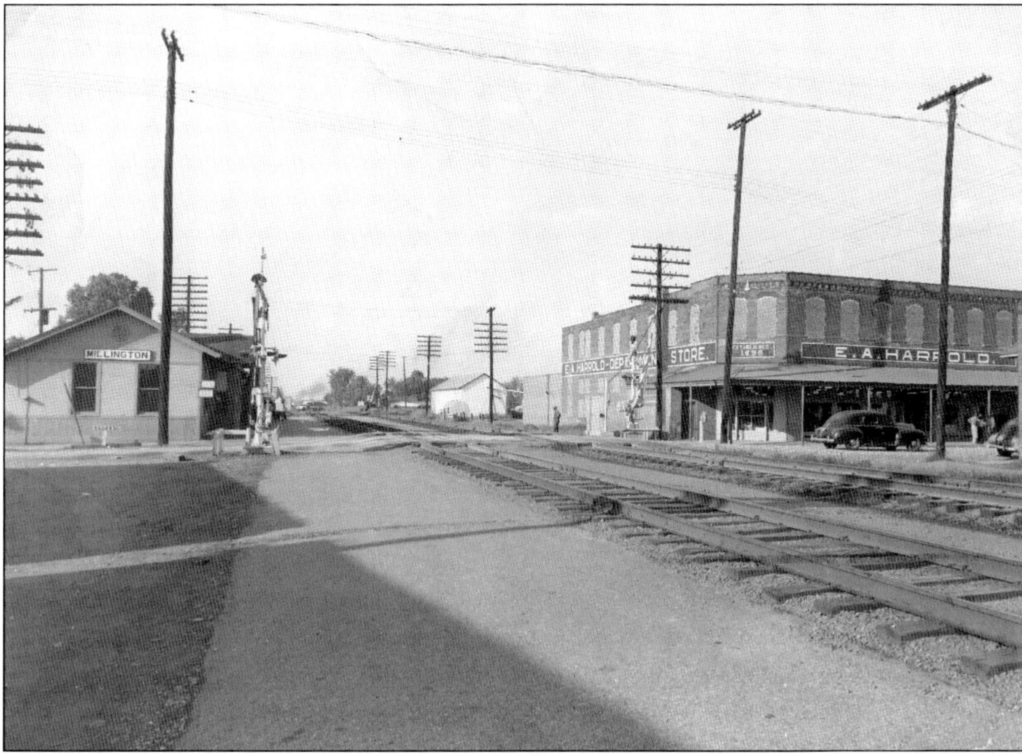

Brickwork on the E.A. Harrold store, built in 1911 to replace an earlier frame structure, reads "Established 1898," but Harrold himself reckoned the business was four years older. He told the Memphis *Commercial Appeal* in 1948 that because of his uncertainty about the company's permanence, "until 1898 I carried my records in my pocket." Shown here in 1949, across the railroad tracks from the Millington train depot, Harrold's store closed in 1960. The building later housed a florist shop, a fabric store, and other businesses. At right, W.S. "Babe" Howard supervised renovations to Millington's oldest landmark after he purchased the Harrold building, now home to Millington Cable Company and Old Timers Restaurant.

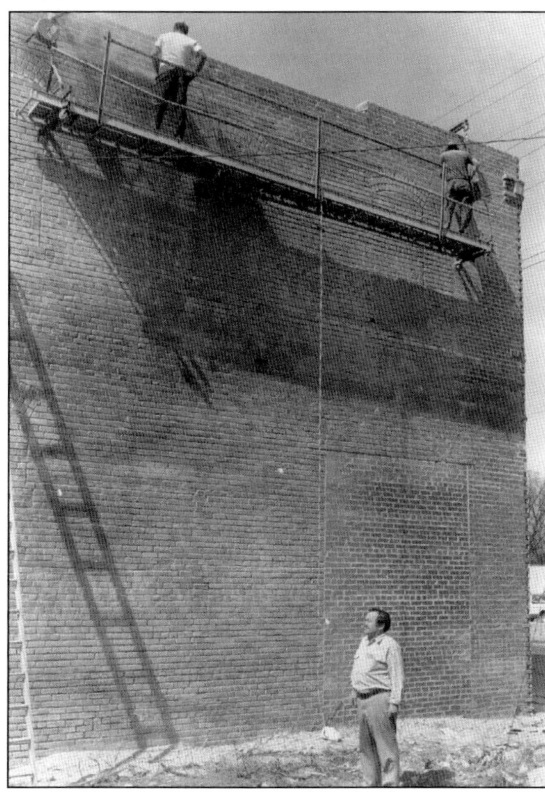

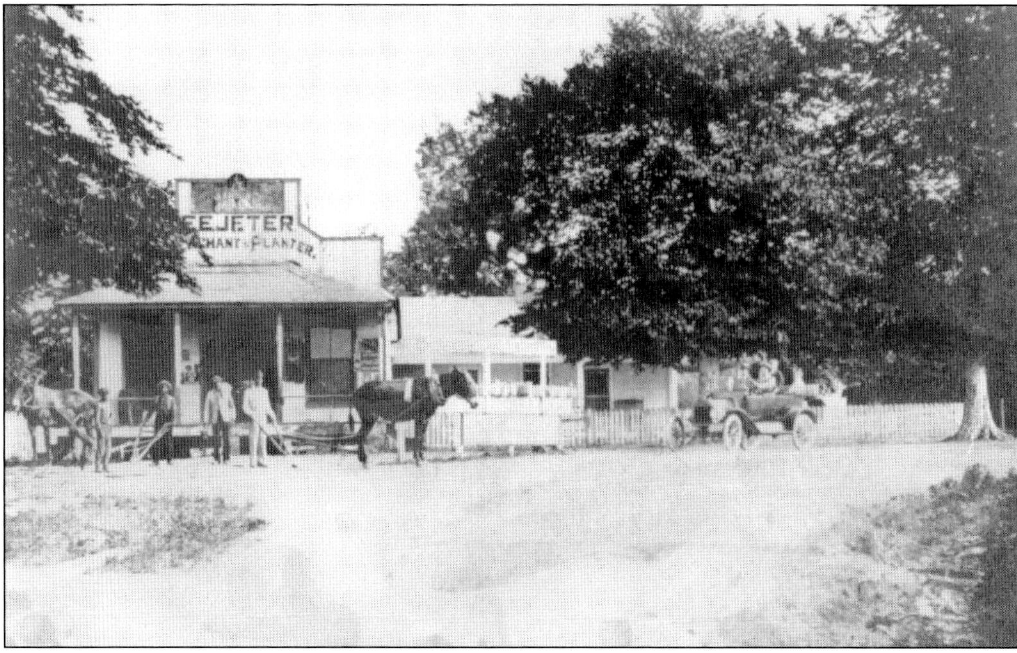

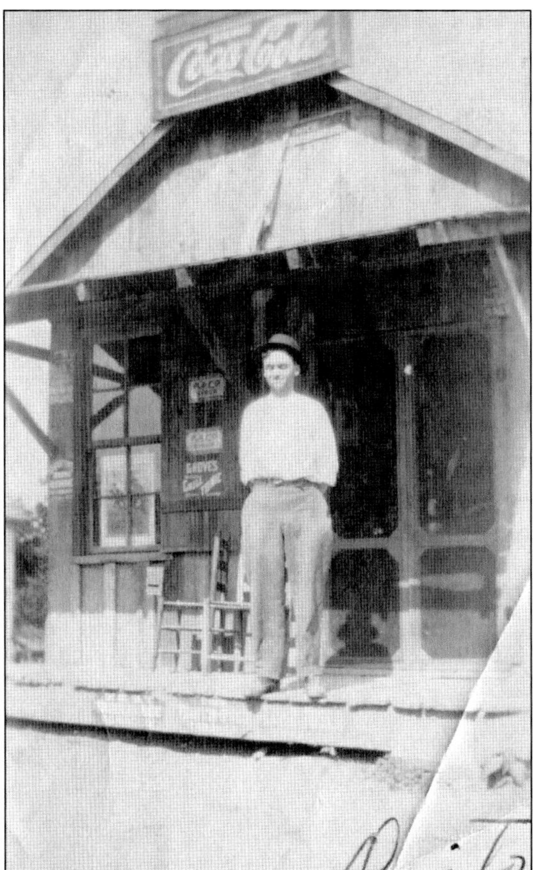

Jeter Store was established at Jackson Hill in 1934 by Emmett Early Jeter and his wife, Dixie, who lived in the house behind the store. A later store, the Shelby Forest General Store, is known for its world-famous cheeseburger and Friday steak nights featuring live bluegrass. It is also a game-checking station, convenience store, bait shop, and local hangout. (Courtesy Doug Ammons, Shelby Forest General Store.)

Aubrey Pouncey stands on the porch of the family store in the Quito community just west of Millington. Pouncey was born in 1895 and was probably no more than 25 in this photograph. The Pouncey Brothers Grocery operated for many decades; in 1971, a burglar entered through a window and stole $3,000 from the store safe. Aubrey Pouncey lived to be 104 years old. (Courtesy MTC.)

E.M. Hicks was a lineman, and his sisters Vivian and Lottye Bartlett were operators for an early telephone company, which offered only day service and had but 40 subscribers. B.L. Howard and his wife, Bessie, both had backgrounds in the telephone industry, and in 1928, they purchased the struggling company from Bass and Hill and founded the Millington Telephone Company. The first company office was a small building once occupied by cobbler Allen Whitaker's Shoe Shop, right. Whitaker offered O'Sullivan's heels and Cat's Paw non-slip soles and promised to make old shoes like new, which he often did for children as they waited and charged the work to their father's tabs. Millington Telephone Company soon moved into a larger building, below. By 1940, Hicks made his living as an electrician and plumber. (Both, courtesy MTC.)

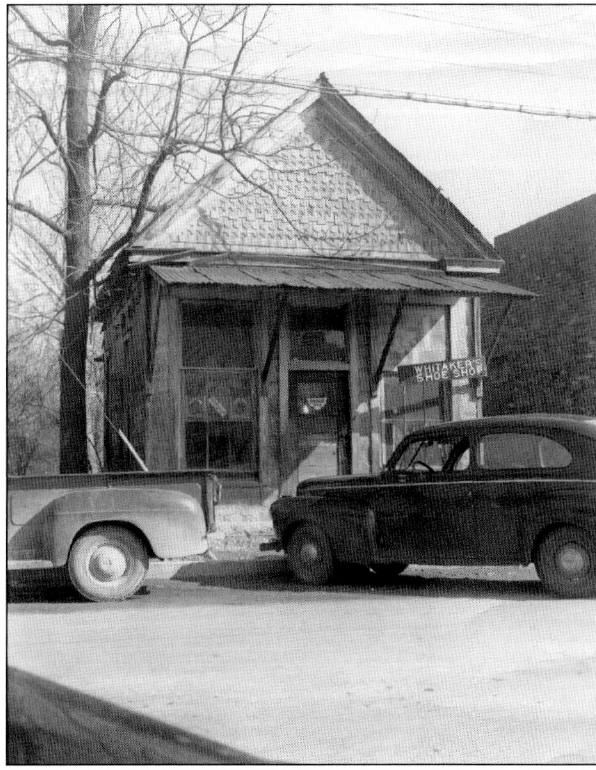

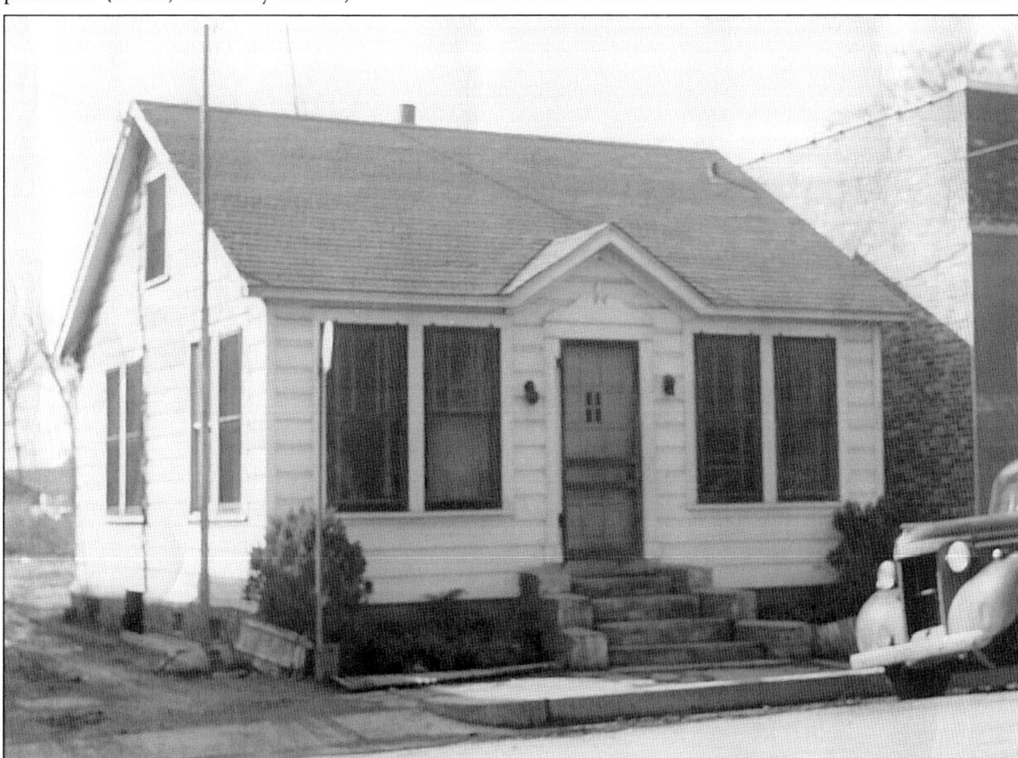

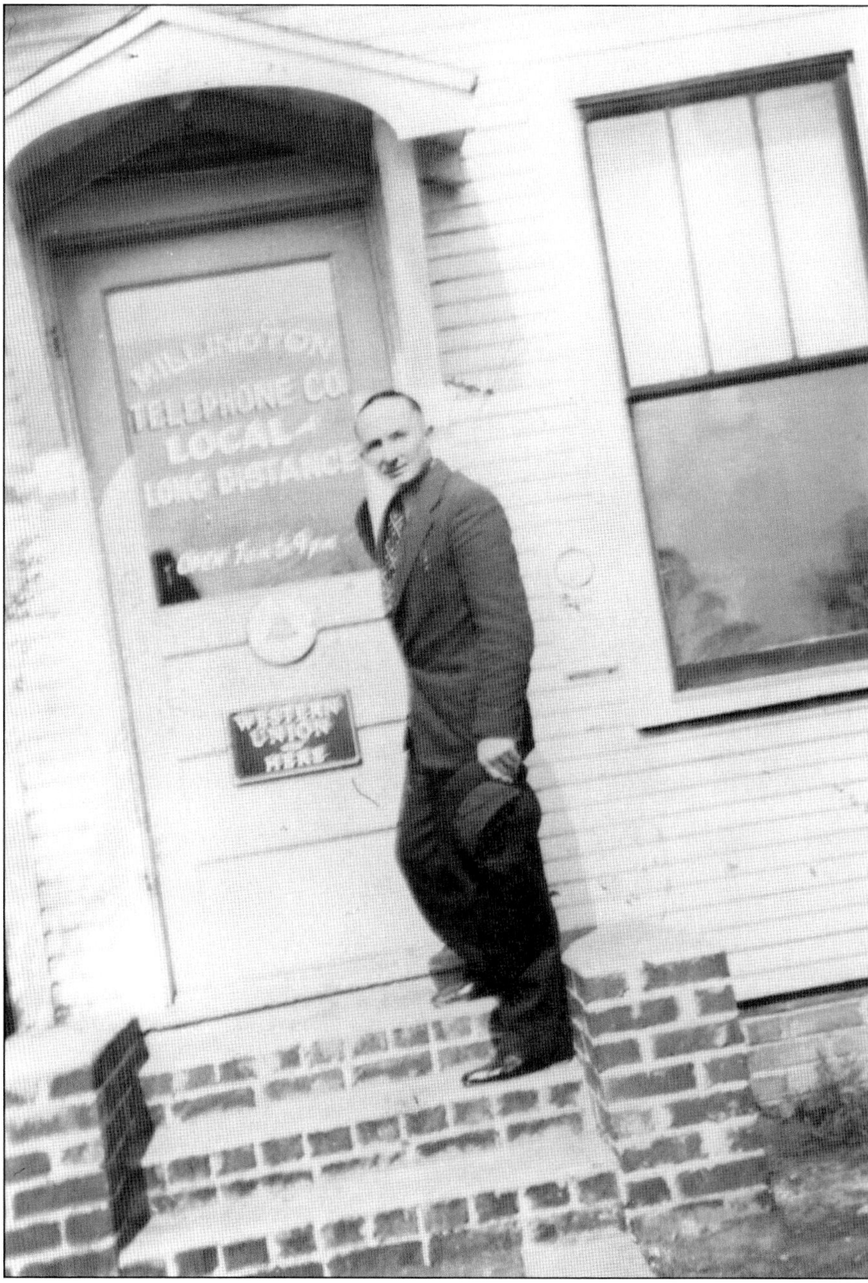

This photograph of B.L. Howard on the steps of the second telephone company office reveals that Howard was attentive to his customer's needs. Renovations to the building included new siding and a cover over a new, more businesslike front door that touted the company's services. These included local and long distance Bell telephone and Western Union services. When the Howards bought the company, service was available only during the day and never on Sundays or holidays and half of the company's 40 subscribers had requested termination of their services. Within two weeks of purchasing the company, Howard expanded to a 24-hour service and he and his wife operated the switchboard in the evenings. By the time this photograph was taken, the office was open 14 hours a day to service over 500 customers. (Courtesy MTC.)

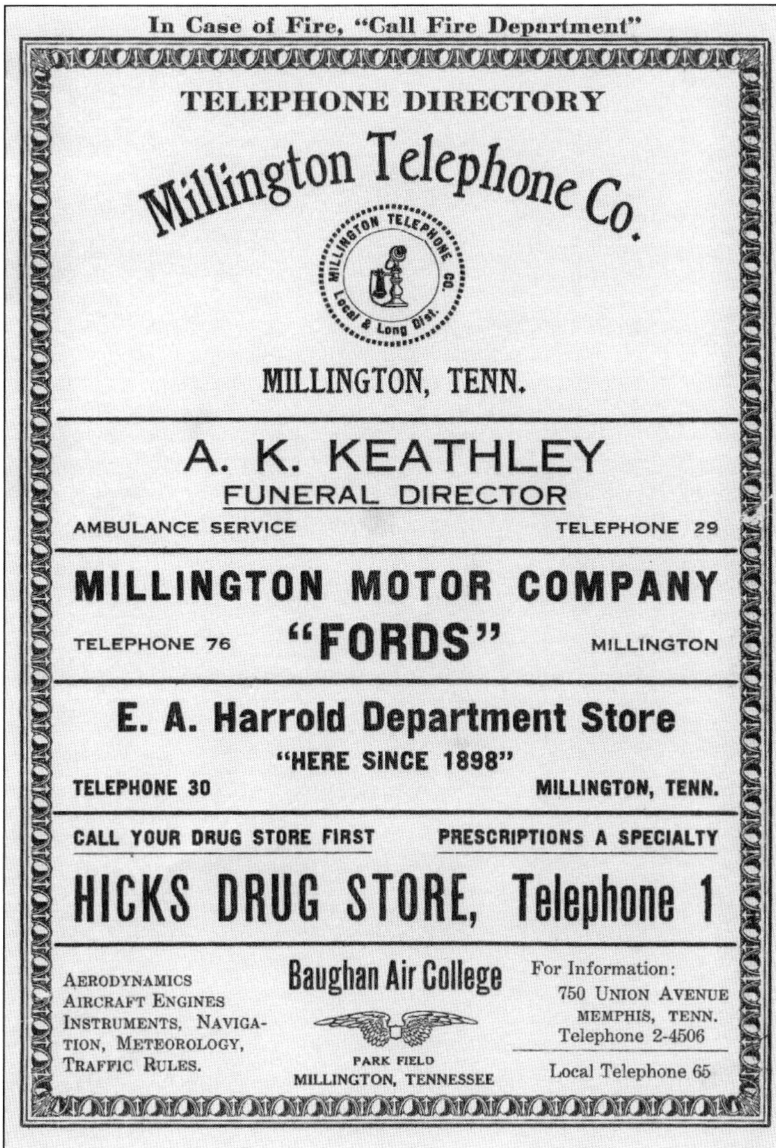

By 1940, Millington Telephone Company was growing rapidly. The telephone directory included 93 local numbers plus three for Dr. H.L. Ellison, who lived and practiced in Memphis and had an office at the Goodwyn Institute. Advertised businesses included Bank of Millington, Baughan Air College at Park Field, Charlie Bower's Restaurant, Ed Cole's meat market and cotton cash store, E.A. Harrold Department Store, E.M. Hicks plumbing and electrical, the W.S. Hornsby Lumber Company, Hotel Crenshaw (rooms were $1, meals 50¢), Grady McMillen's Chevrolet dealership, Millington Motor Company, Pennington Tea Shoppe, Uncle Alex's (Redditt's) Coffee Shop featuring "Aunt Sally's Homemade Pies," Tisdale Garage, Val E. Williams Insurance, and the W. B. "Red" Williams Service Station. Other interesting listings were A.K. Keathley, funeral director and ambulance service; Hicks Drug Store (Telephone 1); and Dr. W.N. Hall, veterinarian and physician. Store listings outside of Millington proper included W.A. Peeke and Bell at Lucy, Williams and Hanks at Kerrville, Thweatt Brothers Store at Shakerag, and the J.N. Harrison store at Brunswick. (Courtesy MTC.)

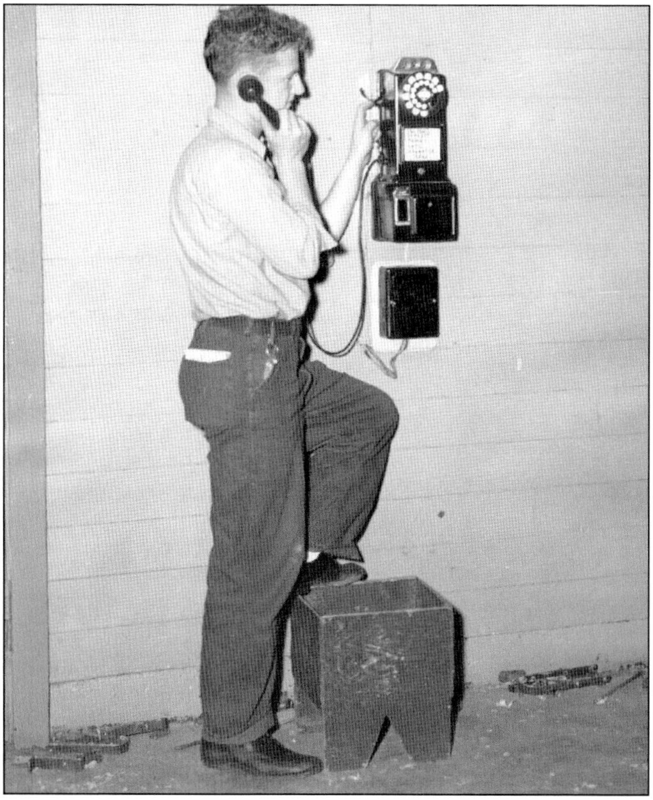

A new Millington Telephone Company building burned to the ground after all but the switchboards had been installed, but like a phoenix, a new central office rose from the ashes and opened for service before the naval base construction was complete. Navy firemen stationed at building N-16, the base fire house, appreciated having a pay phone within convenient reach. (Courtesy MTC.)

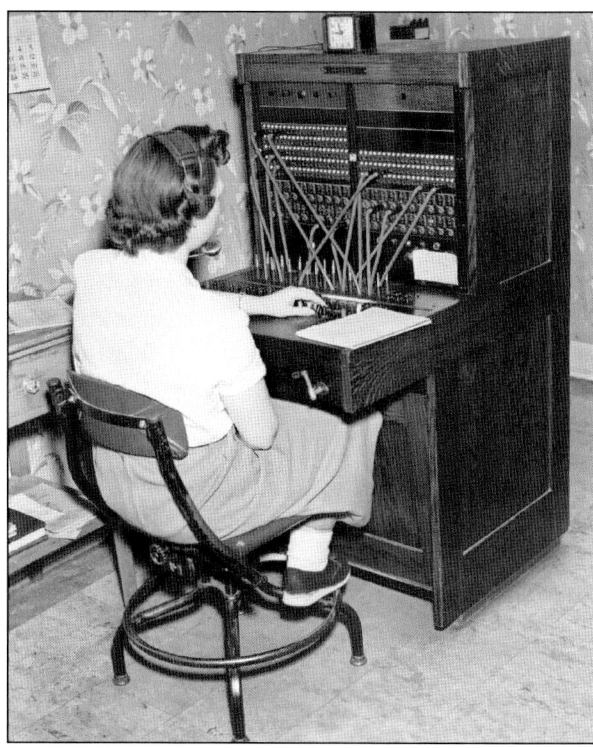

After the war, discharged servicemen taking advantage of the postwar construction boom placed extraordinary demands on telephone companies across the country. A Rural Electrification Administration loan helped Millington Telephone Company make a move toward modernization, but it was five more years before dial service replaced switchboards. Operators in the company's Munford office remained on the switchboard well into the 1950s. (Courtesy MTC.)

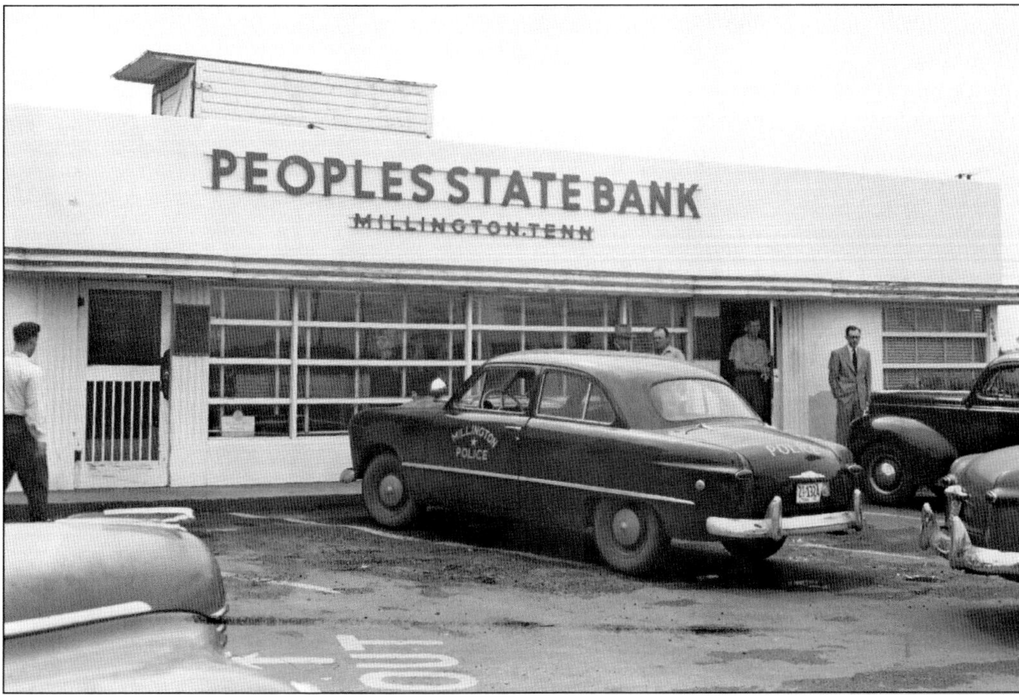

Peoples State Bank (now Trustmark National Bank) opened in 1940. On April 30, 1950, John Wesley Ames, a 20-year-old sailor stationed at Millington's naval air station, used a toy pistol in a daring but unsuccessful stick up. Billy Best of Kerrville saw Ames discard the toy gun in a brush pile as he made his escape. Best turned the toy over to Millington Police.

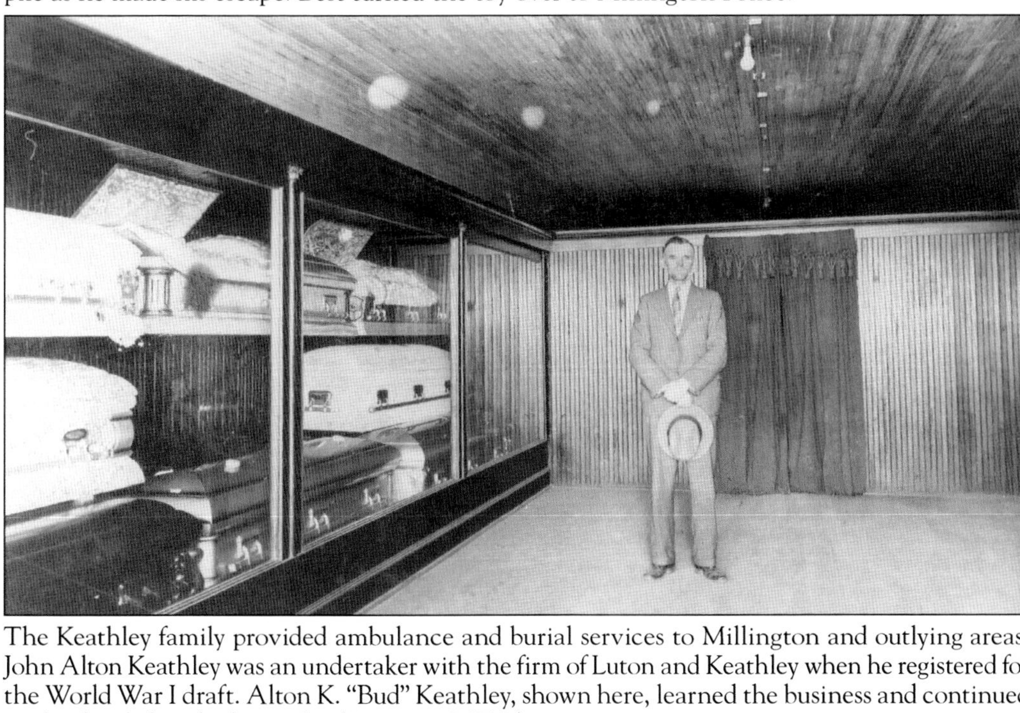

The Keathley family provided ambulance and burial services to Millington and outlying areas. John Alton Keathley was an undertaker with the firm of Luton and Keathley when he registered for the World War I draft. Alton K. "Bud" Keathley, shown here, learned the business and continued it after his father died in 1958. (Courtesy MTC.)

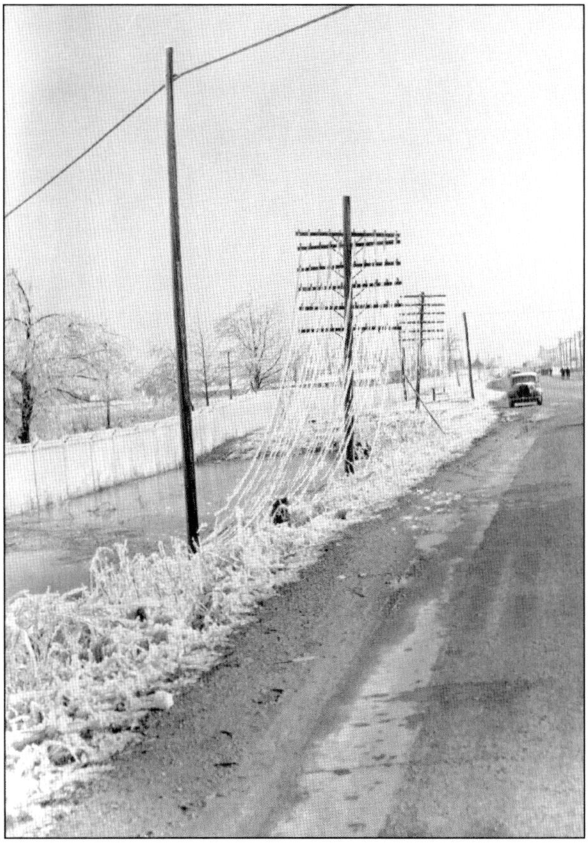

Millington photographer A.O. Conduitte captured these images on January 4, 1951, following the worst sleet storm in local memory. Conduitte's poetic caption for the image to the left read, "A main toll line succumbs to winter's fury along a West Tennessee highway." Below, Millington Telephone Company lineman Leo Holland worked to repair the damages caused by the storm. Damaging sleet storms wreaked havoc on telephone services in 1917 and 1933, when it took six weeks to restore all services. A stronger type of wire was in place by 1936, sparing subscribers from significant outages, and by 1940, two hundred citizens had subscribed. When the 1950 storm hit, only 40 customers still had service, and it took nearly six weeks to restore service to the other 500. (Both, courtesy MTC.)

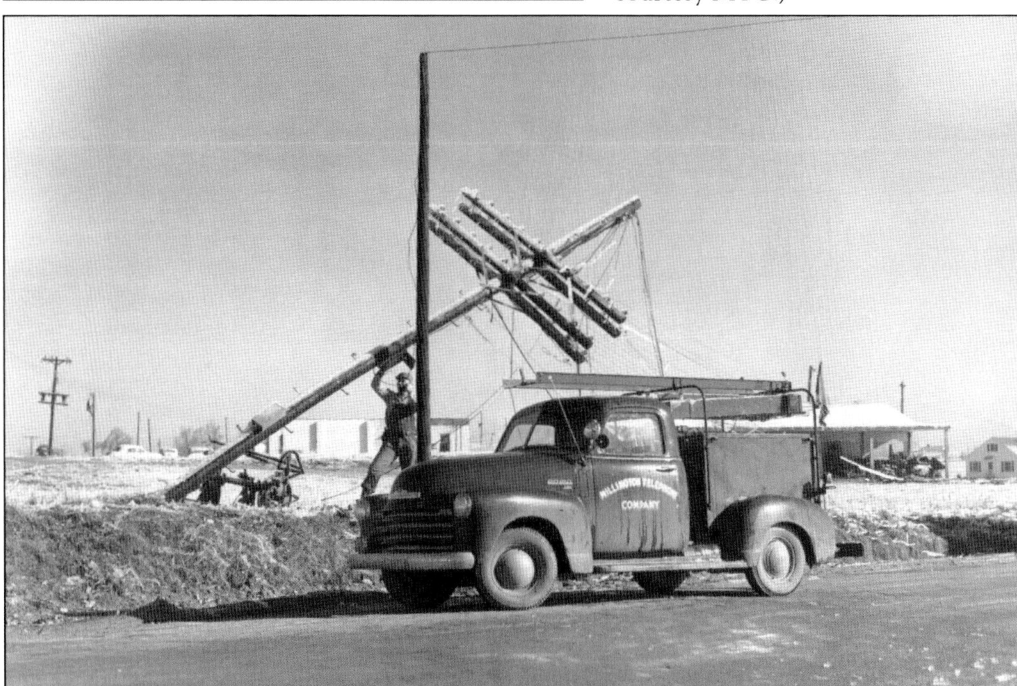

Improving the skills of linemen and ground crews became a priority after the 1951 sleet storm, particularly considering the rapid rate of service expansion in the coverage area. Ground crewmen John Willis and Herdie Swanson, at right, make certain their company truck is adequately stocked for the day's work. The lineman's job has always been a dangerous one, but high demand as Americans grew increasingly dependent upon electrical and communications services made it more so until improvements in safety equipment reduced risk. Among the perils is the danger of electrocution or other incident that would make teamwork essential to get an injured coworker safely to the ground. Below, a crew uses safety harnesses to practice a utility pole rescue. (Both, courtesy MTC.)

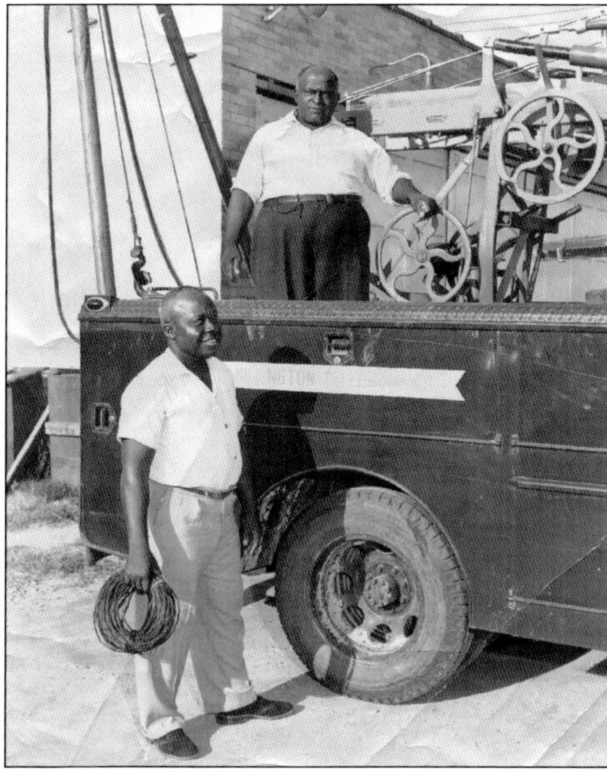

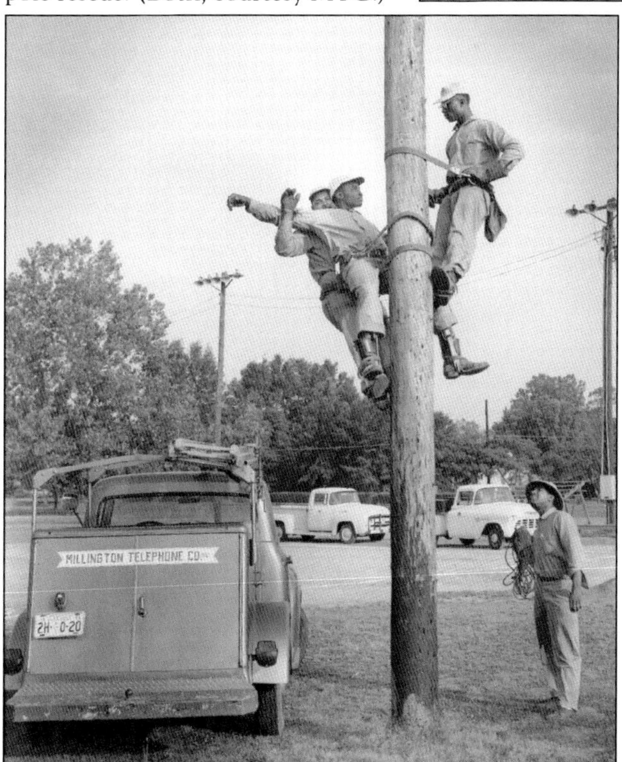

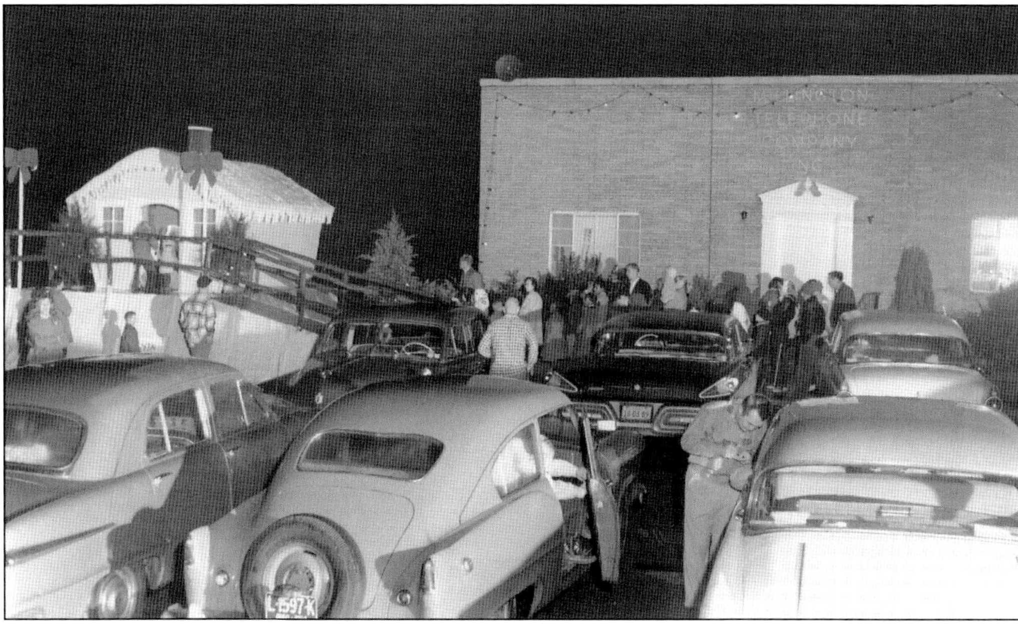

When Millington Telephone Company founder B.L. Howard died in 1953, his son W.S. "Babe" Howard took the helm. Community interaction was always important to the younger Howard. Children especially enjoyed visiting Santa in his cottage on the telephone company grounds. Howard, bottom right, would go from car to car, taking notes from parents on their children's wishes while youngsters waited in line. (Courtesy MTC.)

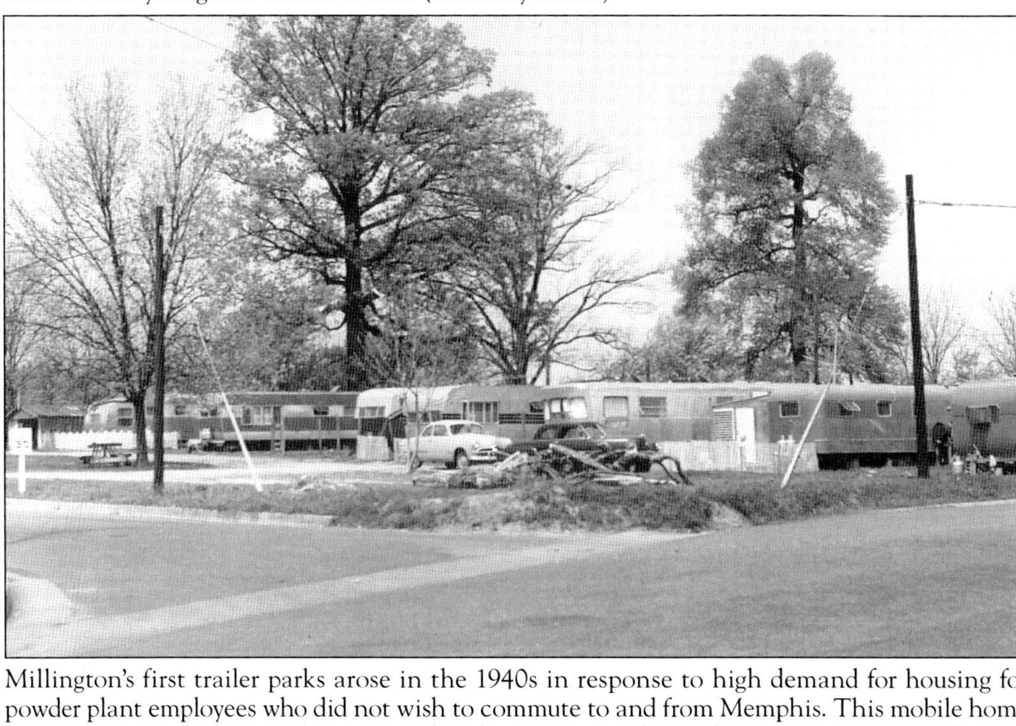

Millington's first trailer parks arose in the 1940s in response to high demand for housing for powder plant employees who did not wish to commute to and from Memphis. This mobile home community remains on Highway 51 and Sigler Lane near the Wal-Mart superstore. Millington Telephone Company installed service lines to the trailer park in April 1954. (Courtesy MTC.)

Millington Telephone Company installed its 1,000th telephone on February 4, 1954. Pictured at right are the honored customer, James E. Parr of Kerrville (seated), installer J.C. Carter, and plant manager Edward Sears (standing). A Millington Telephone Company news release stated, "Mr. James E. Parr is well known in and around Millington and Kerrville where he was a rural mail carrier for many years before his retirement recently. He is now associated with Wages & Whorton, in Millington . . . The Millington Telephone Company, Inc. expects to have approximately 2200 subscribers when it cuts over to dial service during the last part of this year. Many of these subscribers are U.S. Naval personnel and many are rural residents around Millington." Below, a sailor places a call from the pay phone booth located in Building N-2. (Both, courtesy MTC.)

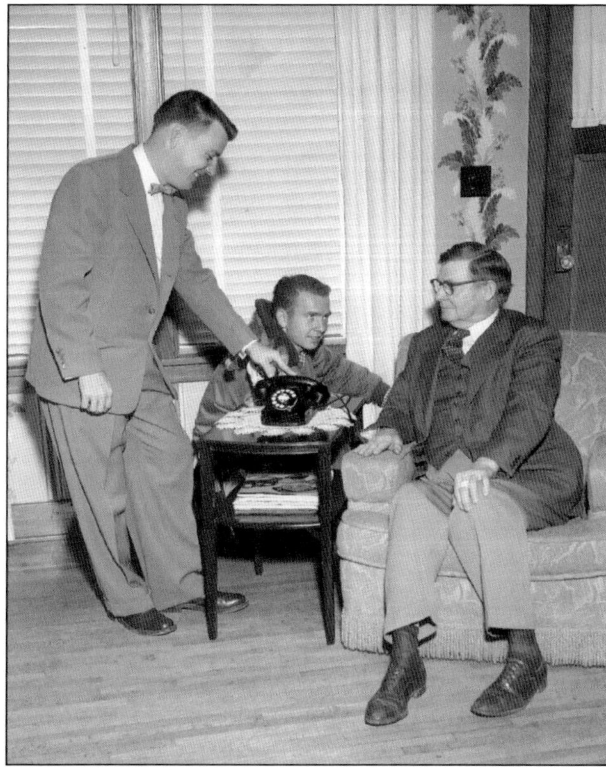

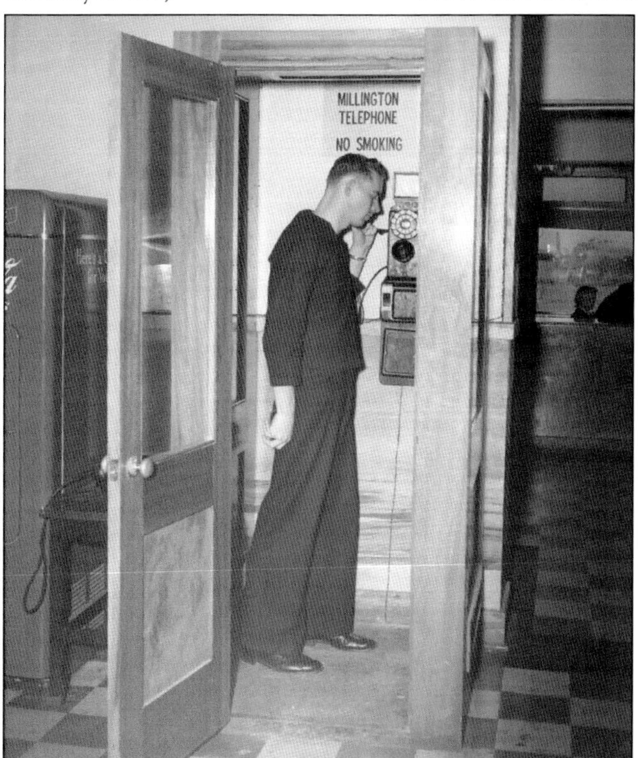

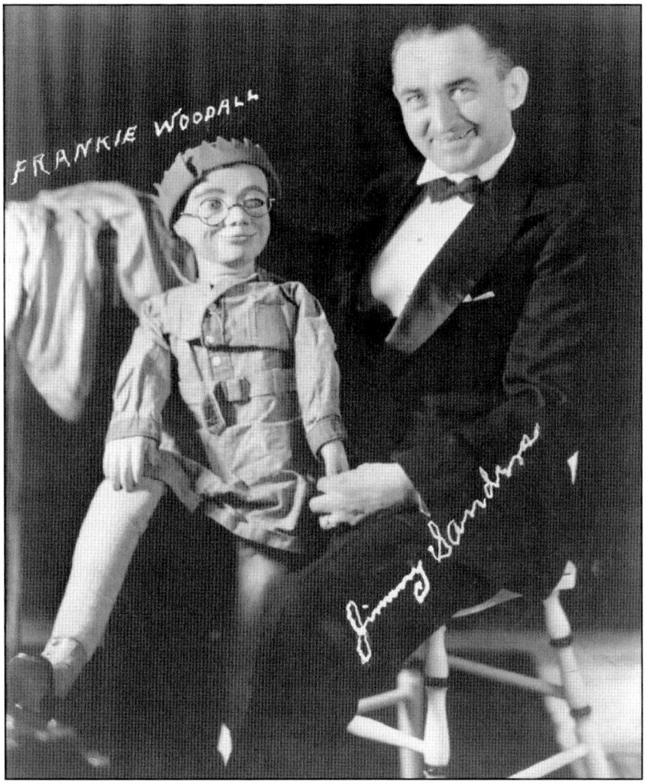

Jimmy Sanders was a Nashville magician known as the "King of Knives." When he visited Millington on October 21, 1954, he brought along his ventriloquist dummy named "Frankie Woodall" to entertain a sizable crowd of telephone company employees and citizens who gathered to celebrate a milestone in the town's history. The Millington Telephone Company Cut-Over barbeque on February 4, 1954, celebrated the company's 1,000th customer and its move to dial service from the original operator switchboard setup. Decorations included floral arrangements in the forms of an oversized telephone and a giant telephone dial. Sponsors included Person Chevrolet Company and Sheffco, a service station located at the corner of Navy Road and Highway 51. (Both, courtesy MTC.)

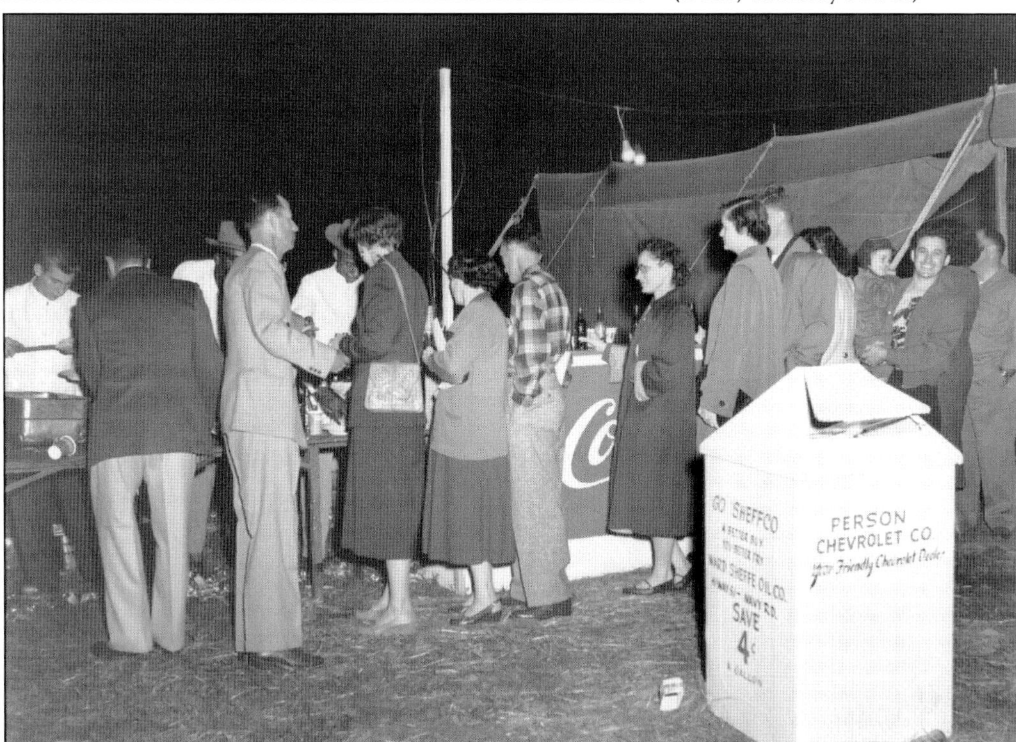

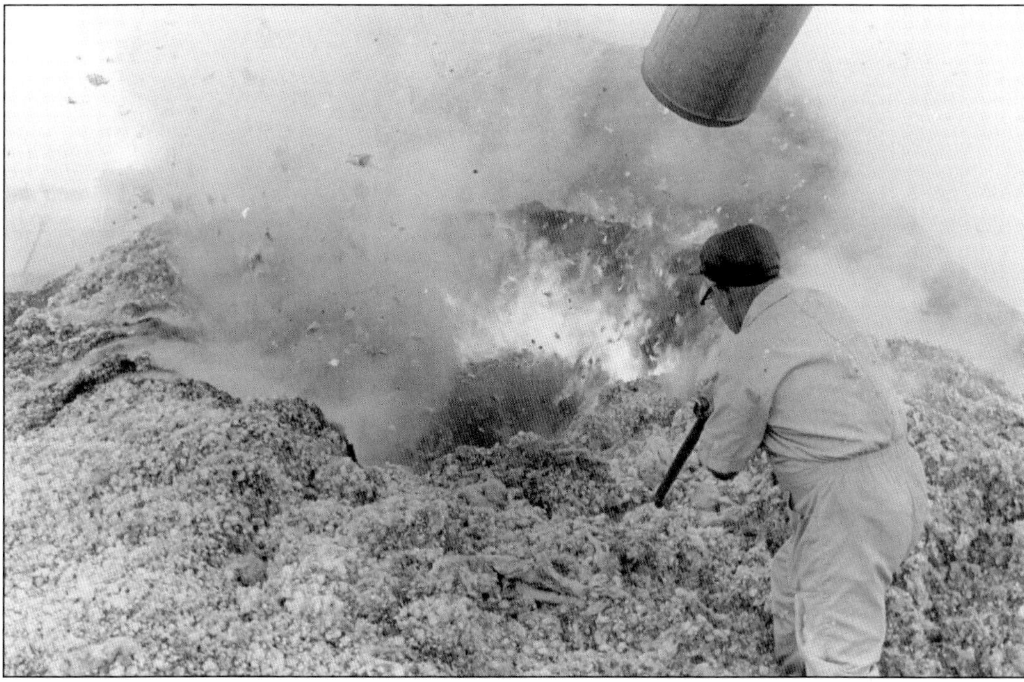

In late October 1963, fire destroyed one Millington cotton gin and threatened another. H.S. "Ham" Mitchell's $75,000 cotton gin was completely consumed by flames. A week later at Armourtown, east of the Navy base, United Gin Company owners Hamp Dobbins, Aaron Longmire, Sidney Ray, and Allen McNamara found their gin, weigh house, seed house, and crops threatened when a fire broke out in dry cotton lint piled behind the gin, as seen above. Below, the Cloverdale Volunteer Fire Department soaked nearby cotton crops, some of which had ignited. Their pumper ran out of water, forcing a return to the station to refill the water tank, but the gin was ultimately spared. The company lost its original gin on the west side of Armour Road to fire in 1952 and rebuilt across the street.

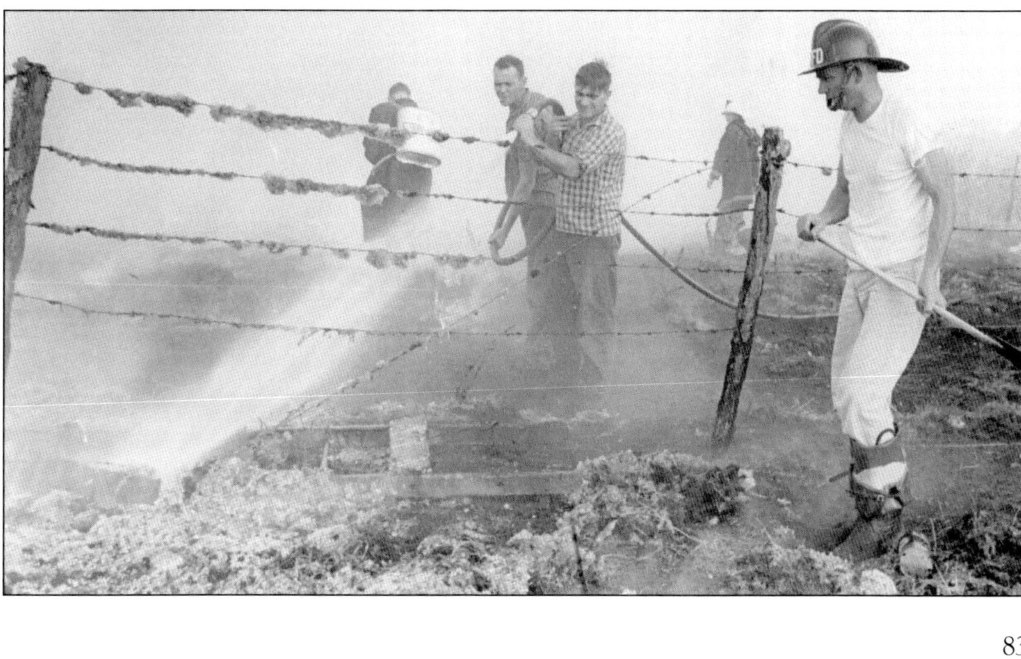

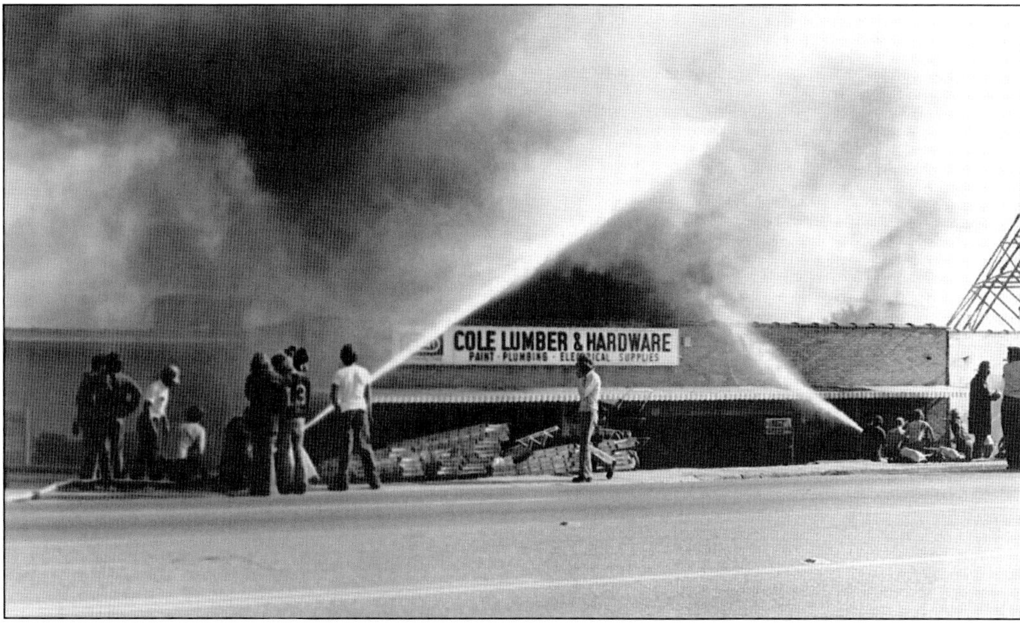

Cole Lumber Company was founded in the 1940s in Drummonds, north of Millington, by sawmill owner Clifton Cole. His four sons incorporated the company as a lumber and hardware business. A branch store opened in Millington in 1963 but burned in 1977. The store was rebuilt and was known as Cole's Do-It-Best Center in 1990. In 2009, the store was replaced with the expanded Cole's Home Solutions.

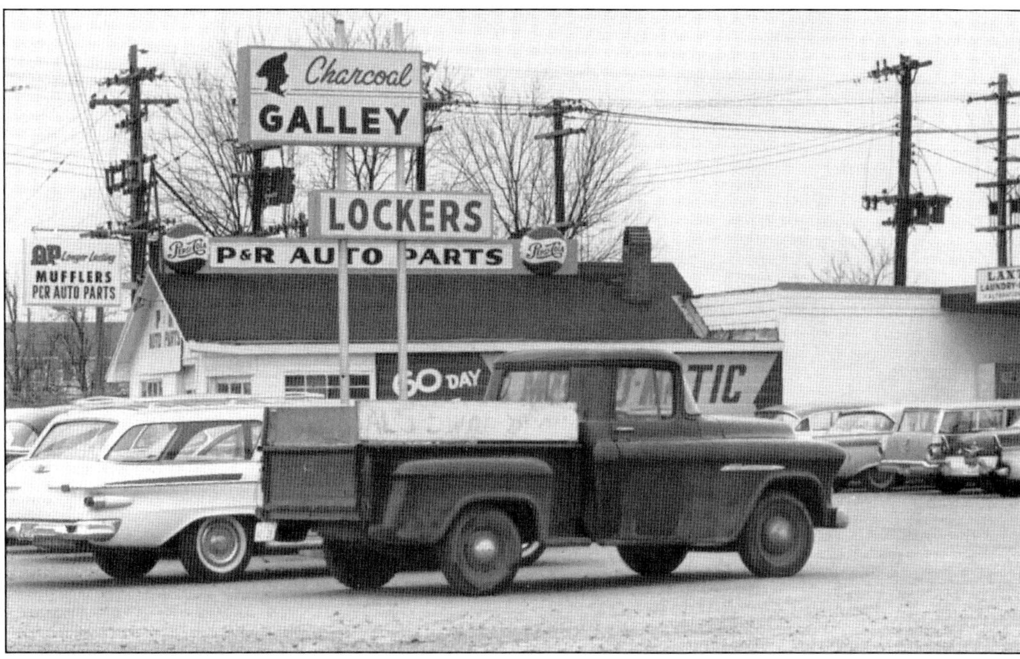

The Jetway Center is located just outside the naval base and has featured businesses that cater to the needs of military personnel for decades. A fire destroyed the original Charcoal Galley in 1963, where many a sailor enjoyed the bottomless coffee for a mere nickel. A locker club rented lockers to military personnel, who were forbidden to have civilian clothes on base.

Three

Millington and the Military

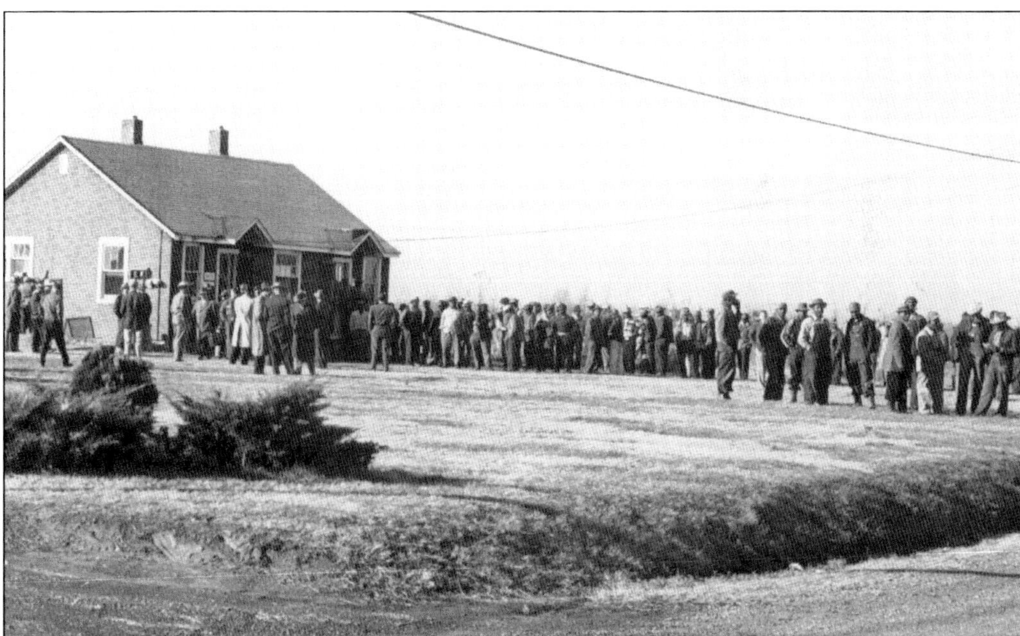

Thousands lined up to apply for jobs at the new munitions factory outside Millington, hoping to directly aid Allied efforts, even before the United States officially entered the war. More than 8,000 were employed on construction crews before the plant opened, and many of those were retained in production jobs.

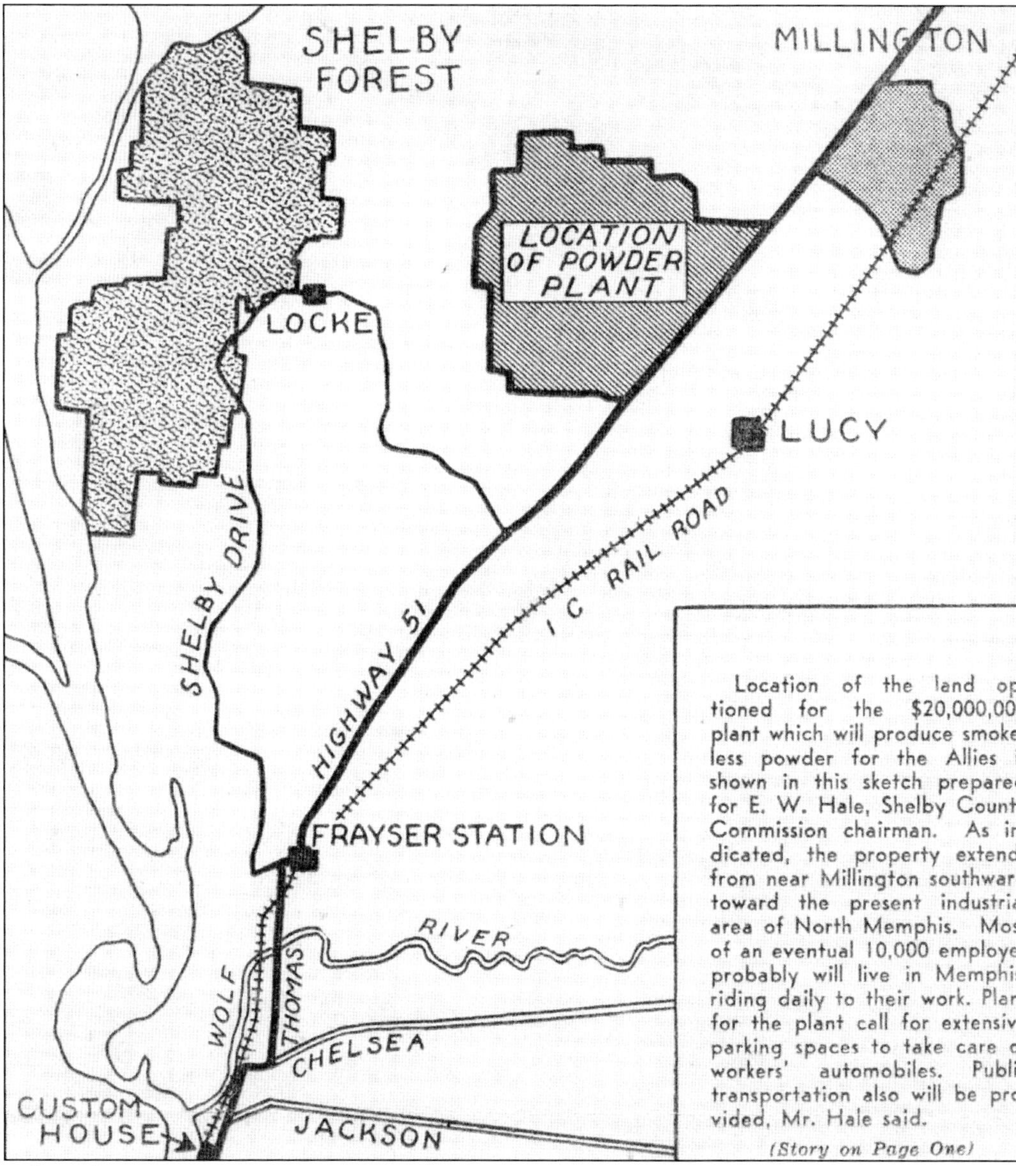

When the Anglo-French Purchasing Board formed the Tennessee Powder Company and commissioned E.I. du Pont de Nemours and Company to construct a $25 million smokeless powder and TNT factory in West Tennessee, few Shelby County residents could have dreamt of its massive scale. As this 1940 map reveals, the Chickasaw Ordnance Works—commonly called the powder plant—was larger than Millington and nearly as large as Shelby Forest Park. A key factor in choosing the site was easy access to cotton linters used in smokeless powder production. The plant, in reality, was a self-contained city. Roughly 6,000 acres were purchased from 70 landowners, including the homes of Millington businessmen E.A. Harrold and William Sigler. The site eventually encompassed 100 buildings, including an electric plant, laboratories, powder magazines, chemical storage, a Pig 'n Whistle cafeteria, police department, hospital, and photographer's studio as well as a two-mile spur of the Illinois Central Railroad. The first properties were bought in July 1940, and production began November 25, 1940. In 1942, the Army took charge of the plant. (Courtesy MPL.)

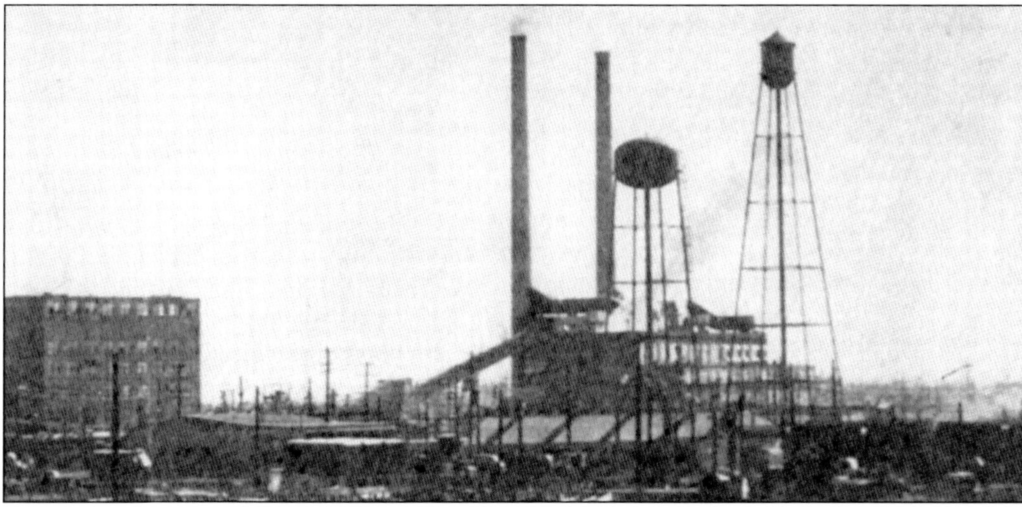

The powder plant drew an estimated 22 million gallons of water daily from eight artesian wells on-site. Area landowners believed their properties were selected for the site because of their locations along the Big Creek Drainage Canal, which ran through the center of the site. Water was stored in a water tower, seen above, and tested hourly, below, for contaminants by chemists Aubert Atkinson (left) and Lacy Keele under the supervision of water engineer James Johnson. The Water Lab was responsible for the purity of drinking water, the safety of water used by employees for washing, and the chemical balance of water used in the power plant and in production. (Both, courtesy MPL.)

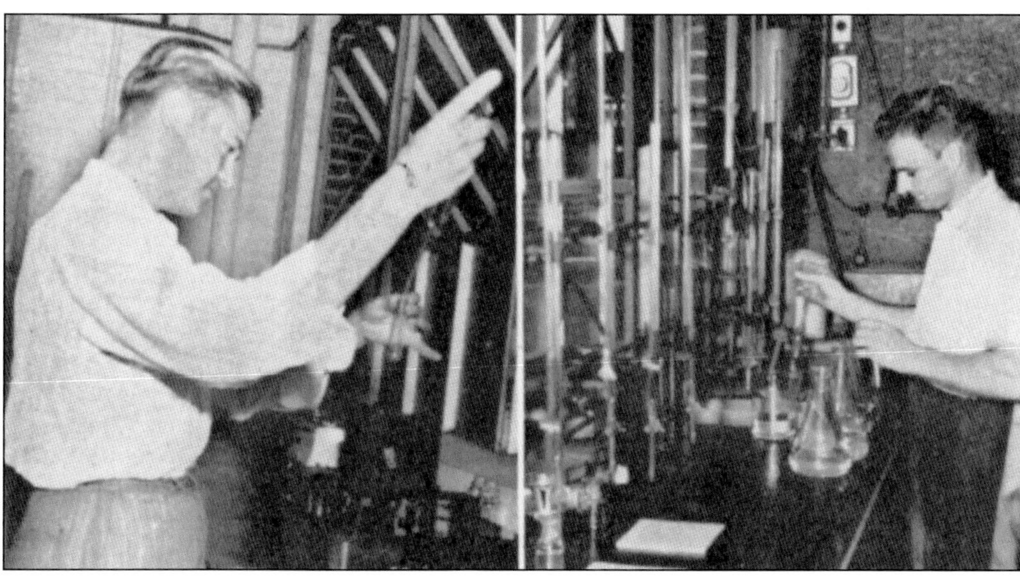

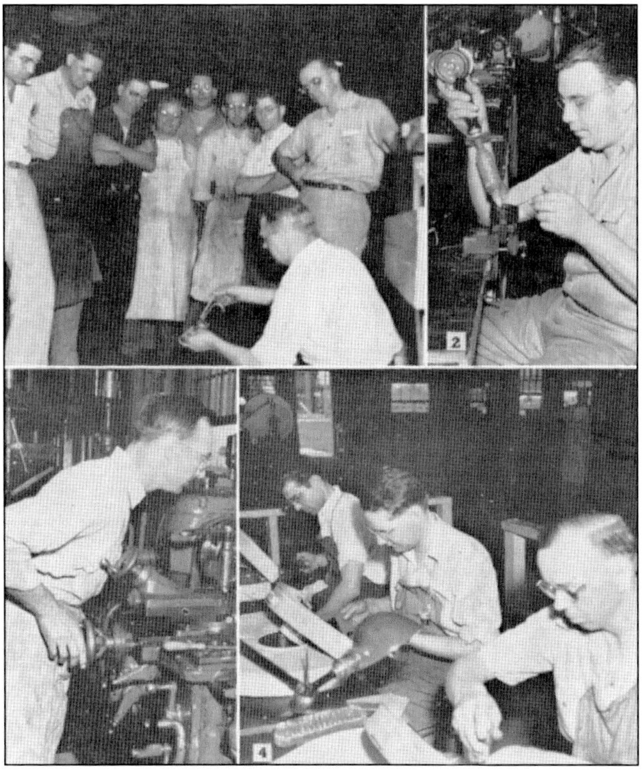

In an era when a favored philosophy was "loose lips sink ships," powder plant employees were sworn to secrecy about their work and few outsiders were allowed on the premises. This collage offers a rare glimpse inside the powder plant. Die shop foreman William Helland (kneeling, top left) discusses the merits of safety goggles with employees. A mechanic uses a soldering iron (top right) while a coworker operates a lathe (lower left). James Gragg, Hugh Hines, and Bill Brooks work in the knife-sharpening area (bottom right). Safety was a major concern; some 7,800 pairs of steel-toed safety shoes were issued to workers and inspected regularly for wear. Below, safety engineer John Meadows inspects finishing area supervisor Fred Redmond's shoes as other employees watch. (Both, courtesy MPL.)

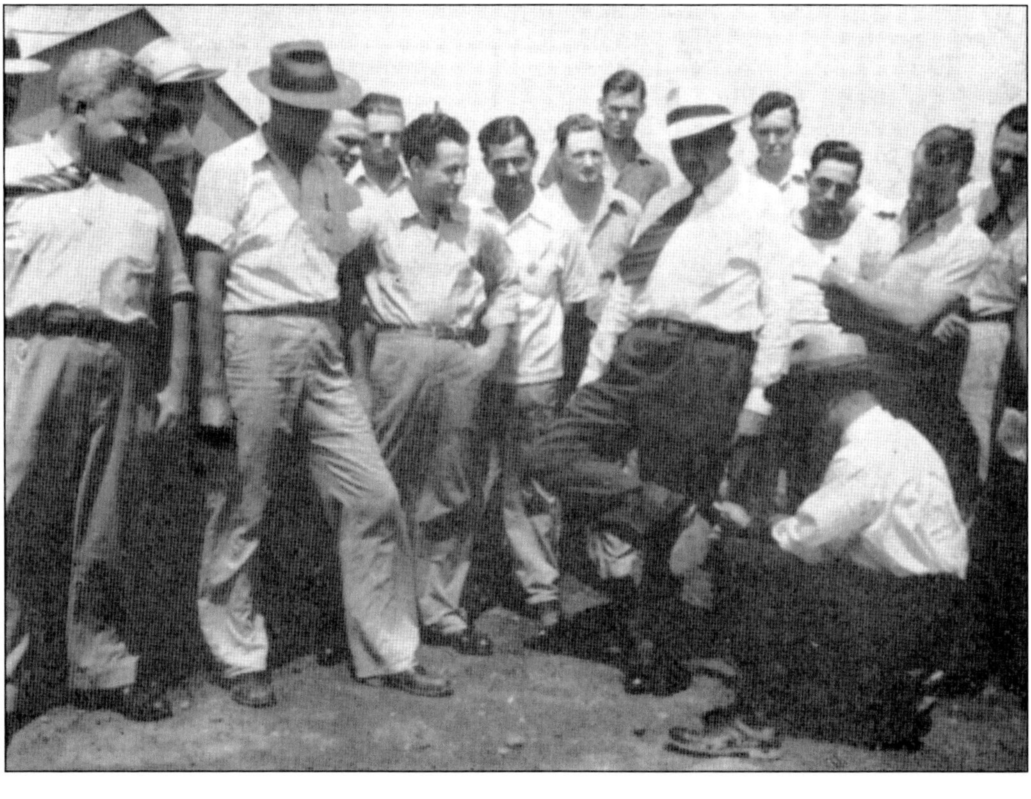

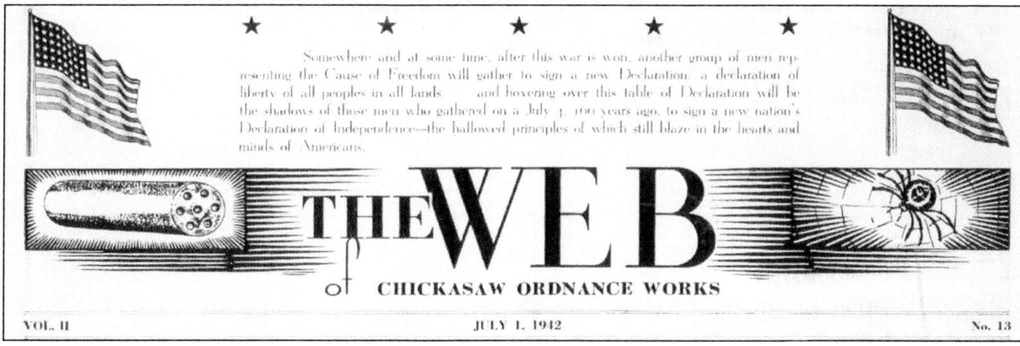
Somewhere and at some time, after this war is won, another group of men representing the Cause of Freedom will gather to sign a new Declaration: a declaration of liberty of all peoples in all lands and hovering over this table of Declaration will be the shadows of those men who gathered on a July 4, 166 years ago, to sign a new nation's Declaration of Independence—the hallowed principles of which still blaze in the hearts and minds of Americans.

THE WEB of CHICKASAW ORDNANCE WORKS

VOL. II | JULY 1, 1942 | No. 13

Former employees of the powder plant are likely to recall its newsletter, the *Web*. The newsletter featured columns on security and safety on the job, on the road, and at home. It also included humor and the latest gossip as well as news of births, illnesses, deaths, and employee military duties. In addition, the *Web* encouraged the purchase of war bonds and participation in supply drives. (Courtesy MPL.)

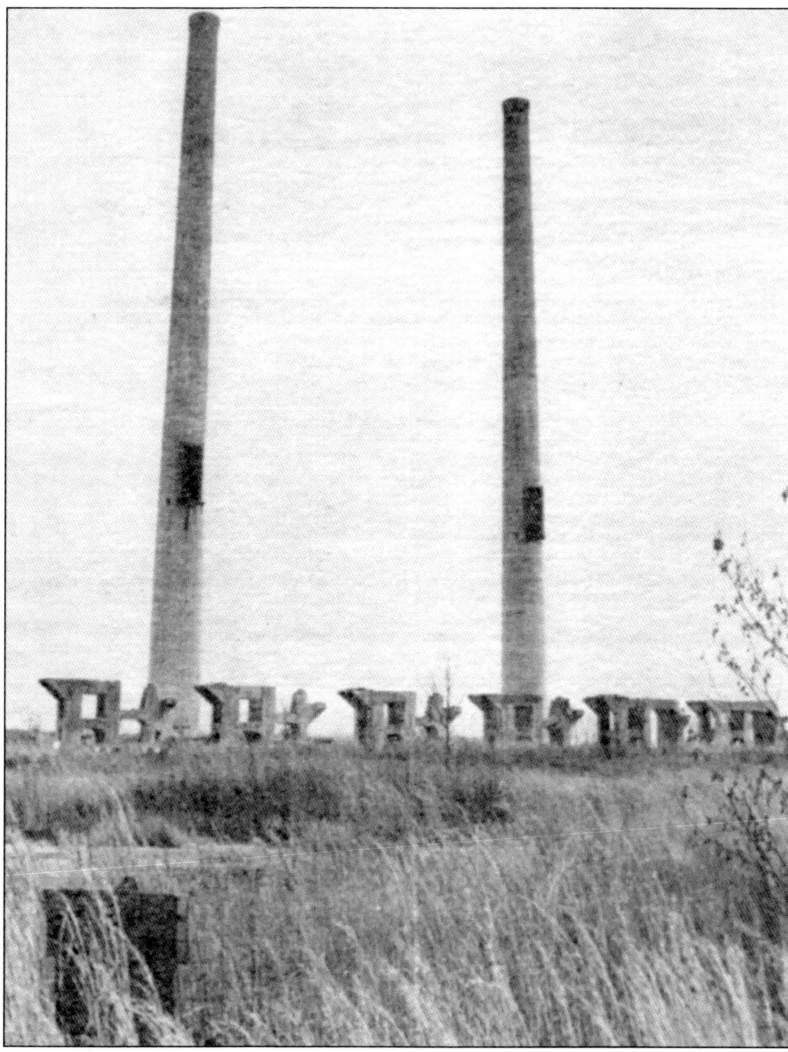

After the war, demolition of the powder plant occurred rapidly. In 1946, over 100,000 pounds of unused powder were burned to avoid long-term storage expenses and plant buildings were dismantled. Three generations have grown up since with nothing more than barbed wire fencing, some concrete bunkers, these smokestacks, and the stories of their elders to prove the existence of this once-bustling wartime factory on the old Shelby Drive. (Courtesy MPL.)

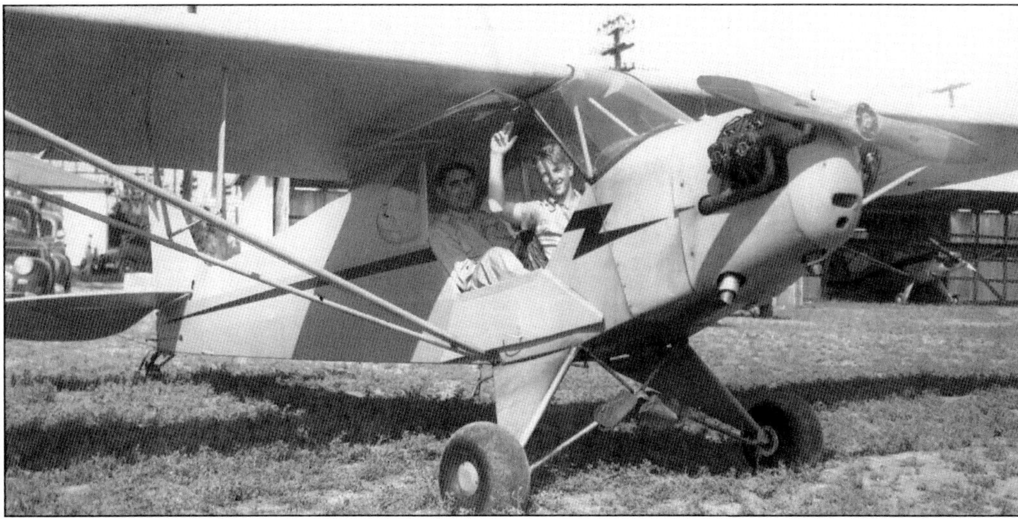

East of Millington in the 1940s, the naval air station was built over Park Field, which had been used for flight training during World War I. Above in 1945, junior aviator Rolland Anderson, one of many youngsters to learn the basics of flight at Millington's naval air station in the postwar years, salutes the camera. Some went on to military service or became commercial pilots, while others put their skills to work operating crop dusters. Later, Boy Scouts acquired skills in modern flight simulators, seen below. (Below, courtesy *Millington Star.*)

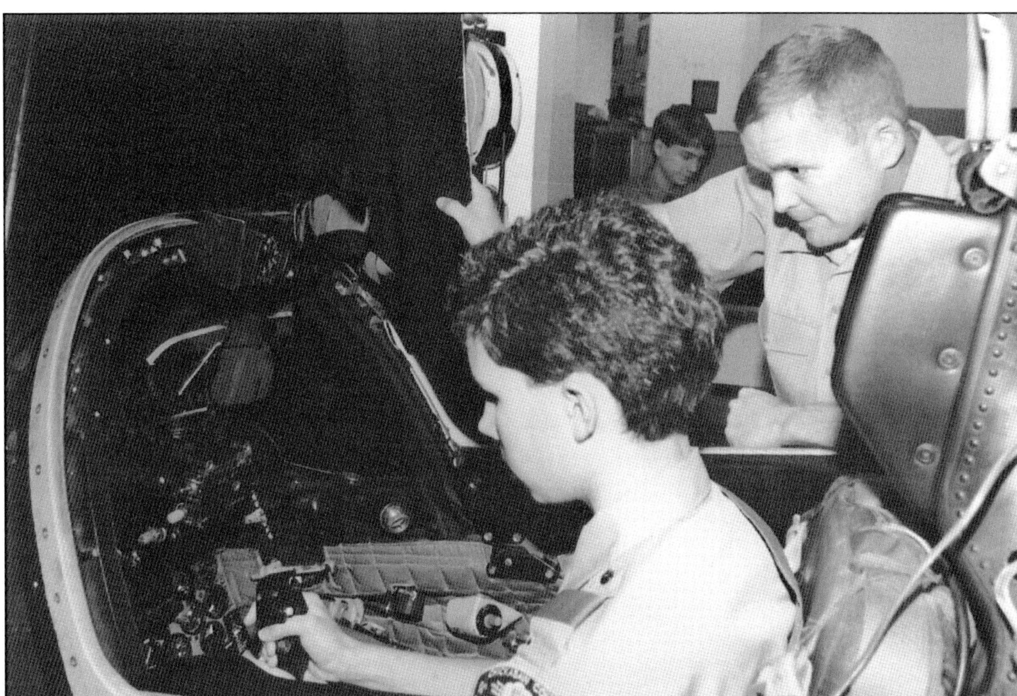

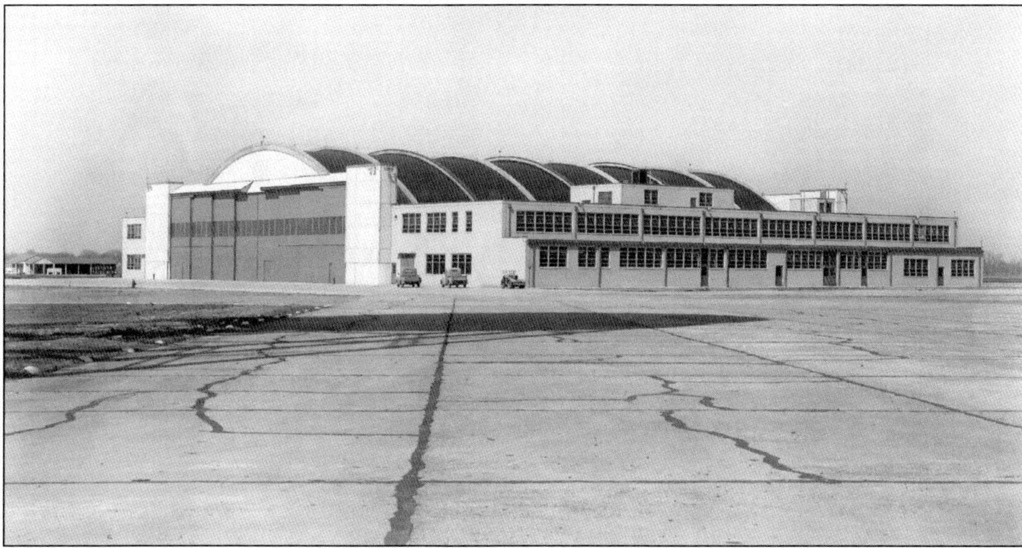

The 1955 replica of a hangar at Denver's Buckley Field, seen above, was the largest building on base at 52 feet high, 150 feet wide, and 241 feet long. It boasted air-conditioning and electric doors 25 cars wide with safety stops similar to those in today's elevator doors. The 1994 Base Realignment and Closure committee moved flight training to Pensacola and made Millington a support operation. The airfield became Millington Regional Jetport. Below, an aerial photograph captures a Navy jet in flight above the air station hangars. (Below, courtesy MTC.)

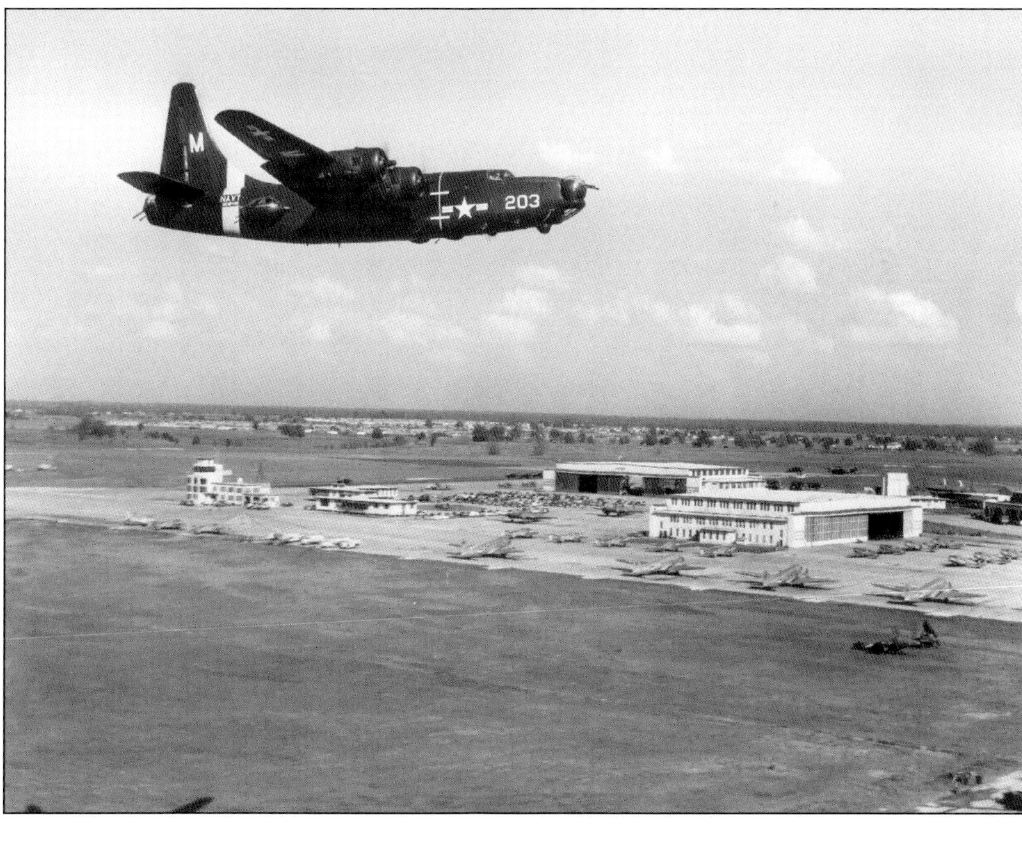

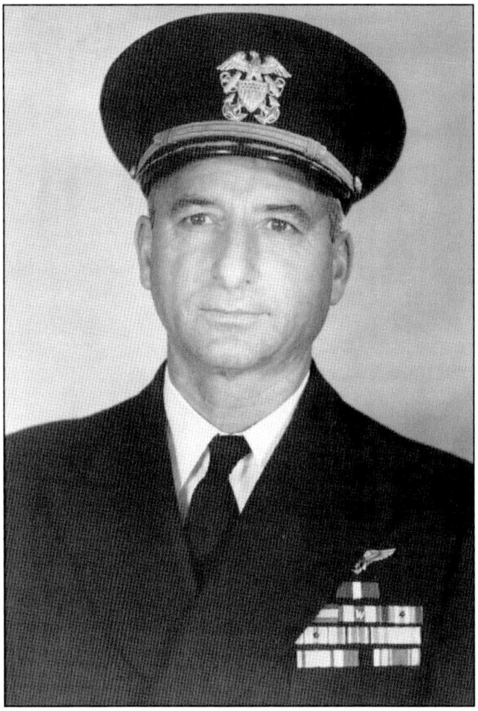

Rear Adm. Winfield Scott Cunningham, pictured here in 1950, was responsible for the first US victory of World War II. Appointed officer in charge of all naval activities at Wake Island in December 1941, Cunningham soon became a Japanese POW. He escaped but was recaptured twice. In 1947, he was assigned to command of the Naval Air Technical Training Center at Millington. Although the photograph below was taken a few years before Cunningham arrived, military meals had undoubtedly stayed much the same. Memphis mayor Wyeth Chandler visited the base in 1944 and ate lunch with the troops. (Below, courtesy MPL.)

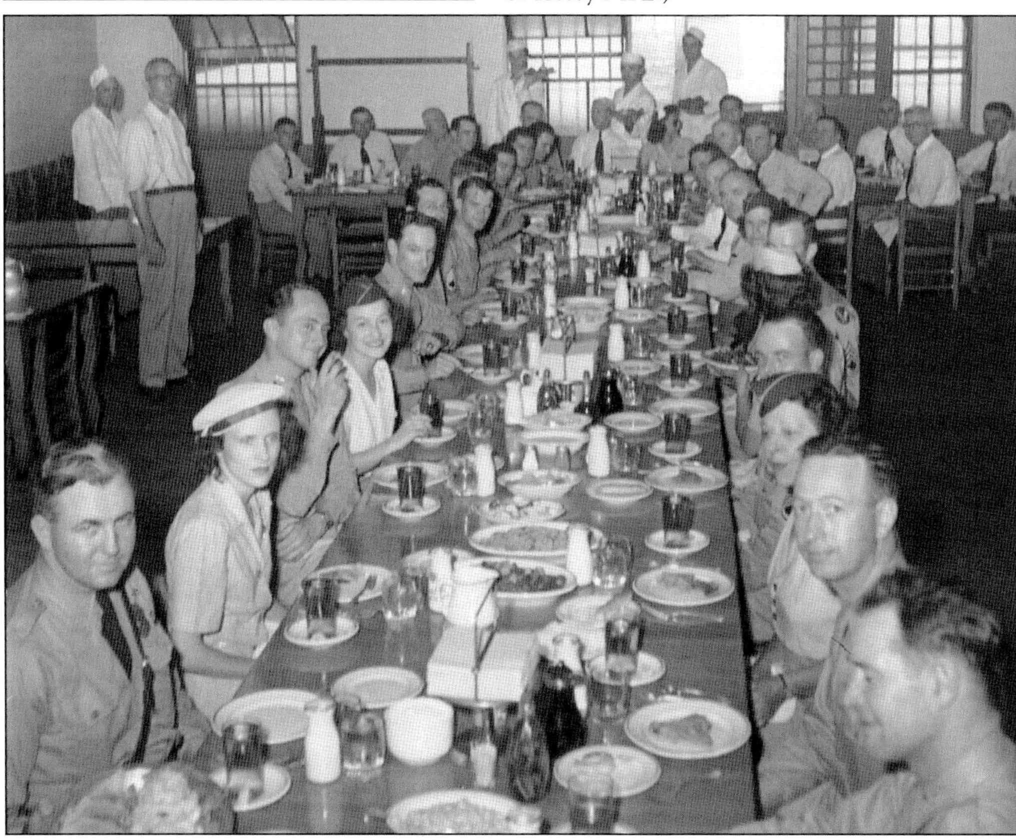

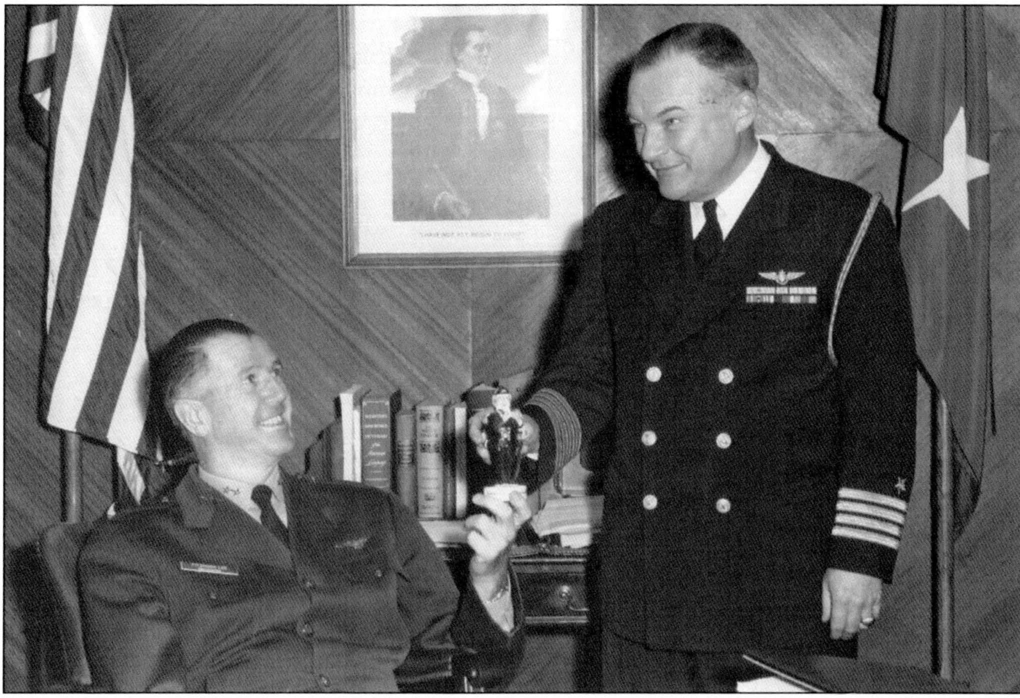

The military has always maintained respect for the agricultural character of north Shelby County. According to the Navy, the "Cotton Pickin' Commodore" was the symbol of naval authority at the Millington base. In February 1960, Capt. B.E. Close, above right, took over as chief of naval air technical training (CNATECHTRA), relieving Rear Adm. Fitzhugh Lee. (Official US Navy photograph; courtesy UM.)

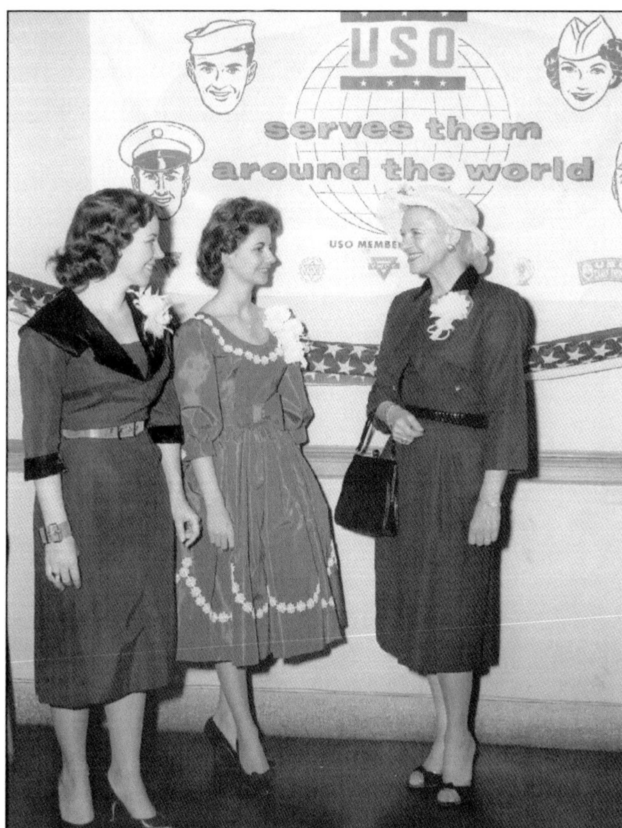

Ladies of the USO have long been avid supporters of Millington's military personnel. Pictured at right in 1960 are, from left to right, Evelyn Byrne, vice president of the Junior Volunteer Council; Beverly Lyons, president of the Junior Volunteer Council of the Memphis USO; and Mrs. B.E. Close, chairman of the Volunteer Workers and wife of the "Cotton Pickin' Commodore." (Official US Navy photograph; courtesy UM.)

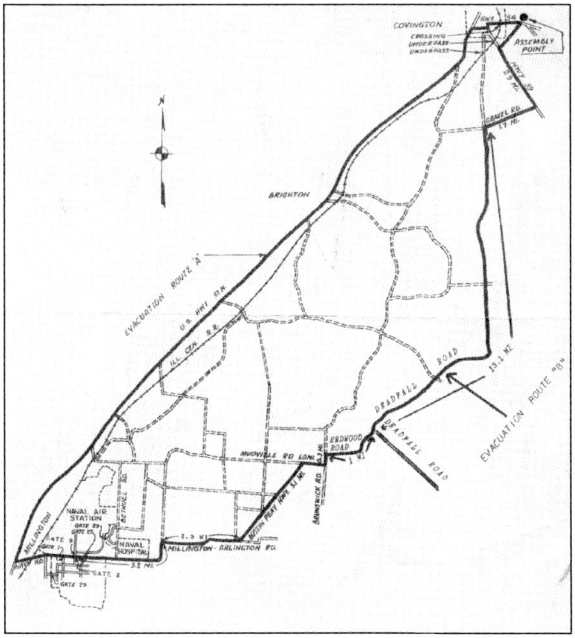

During the Cold War era, schoolchildren everywhere learned the meaning of "duck and cover." In Millington, their parents were advised of officially designated escape routes in the event of an air strike at the naval base. This evacuation map, dated October 1961, indicated the primary and secondary routes to Covington, the Tipton County seat, some 25 miles north of Millington.

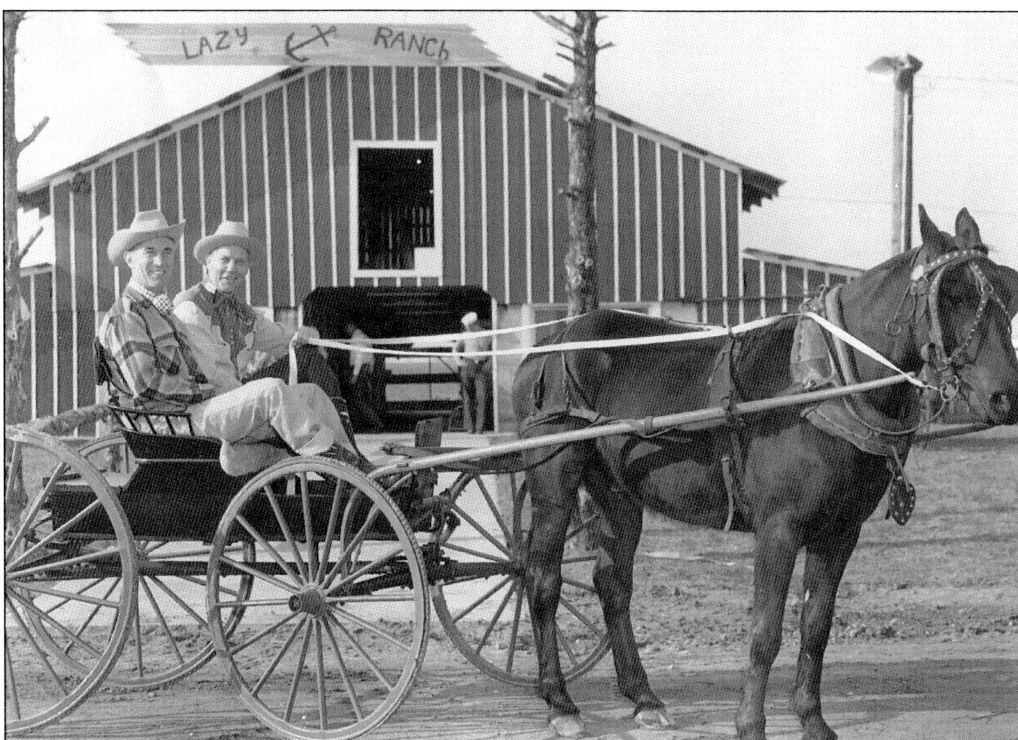

In 1962, the Navy opened the Lazy Anchor Ranch, allowing military personnel, their dependents, and guests to ride quarter horses or Tennessee walking horses over nine miles of bridle paths, see burros, or be carried in one of two buggies or a covered wagon. Pictured on opening day are commanding officer Capt. J.J. Munson (left) and chief of naval air technical training Rear Adm. J.C. Clifton.

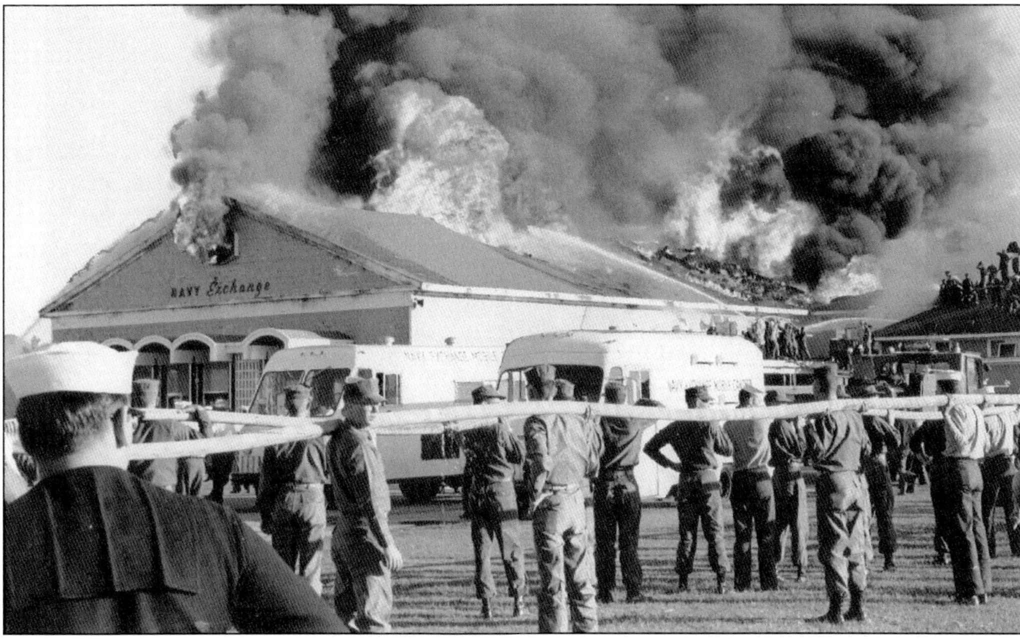

Two fires occurred on the base in one week in October 1966, causing significant damage to several buildings. The first burned the Navy Exchange, which served a function similar to today's Wal-Mart superstores for military personnel. Officers wasted no opportunity for exercise and teamwork, as indicated by the interesting means of keeping fire hoses from becoming tangled. The second fire destroyed two World War II–era enlisted men's clubs, a penny arcade, a photography studio, and a beauty shop.

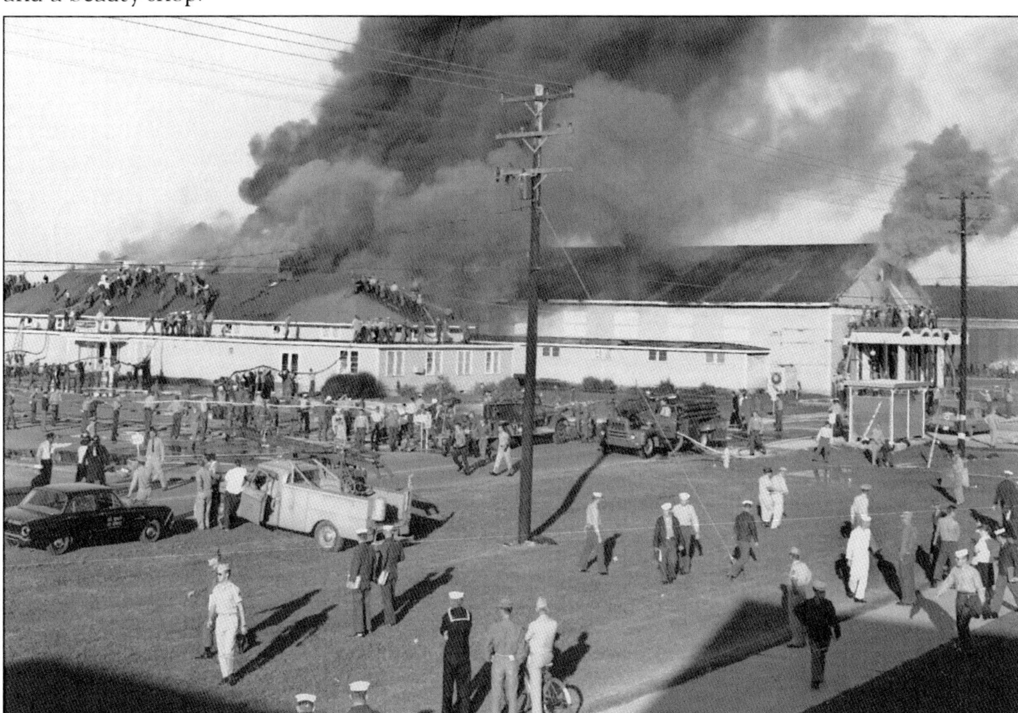

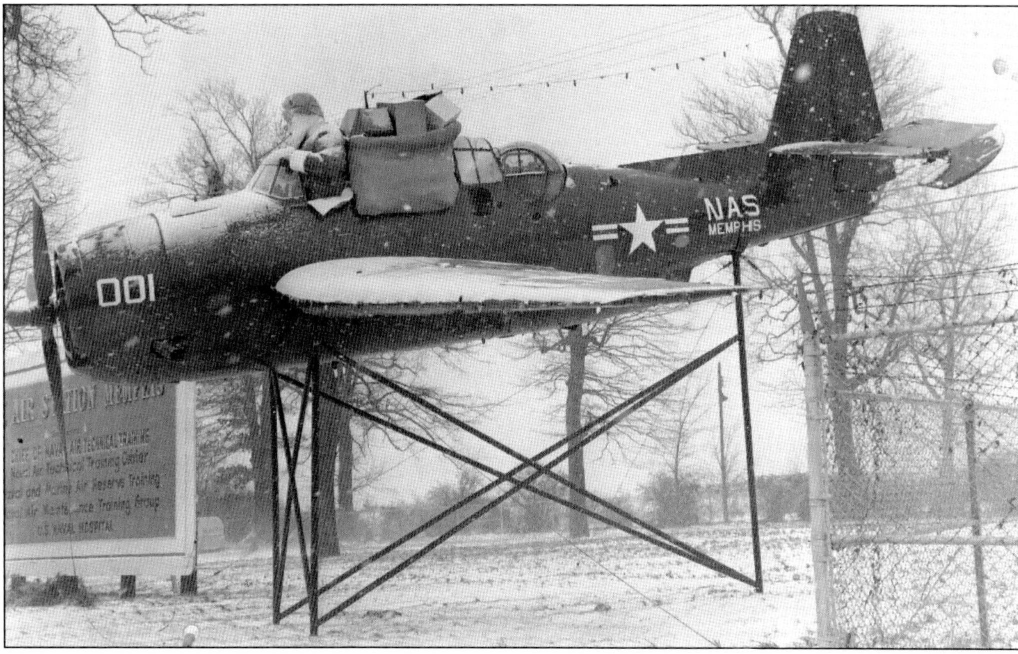

Although often stationed far from home, neither sailors nor Marines went unaware of the Christmas holidays. The above image of Santa Claus using an airplane rather than a sleigh greeted travelers along Navy Road on Christmas Eve in 1966, bringing some jolly good cheer to town. Below, a group of Marines was caught on camera rappelling out of a helicopter. (Below, courtesy *Millington Star.*)

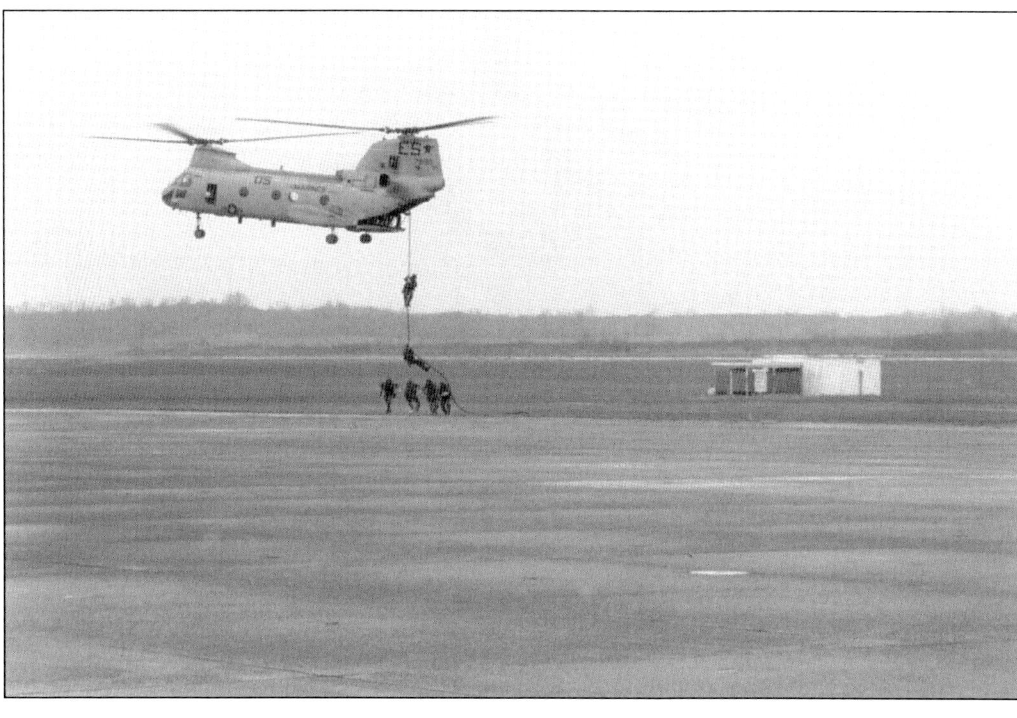

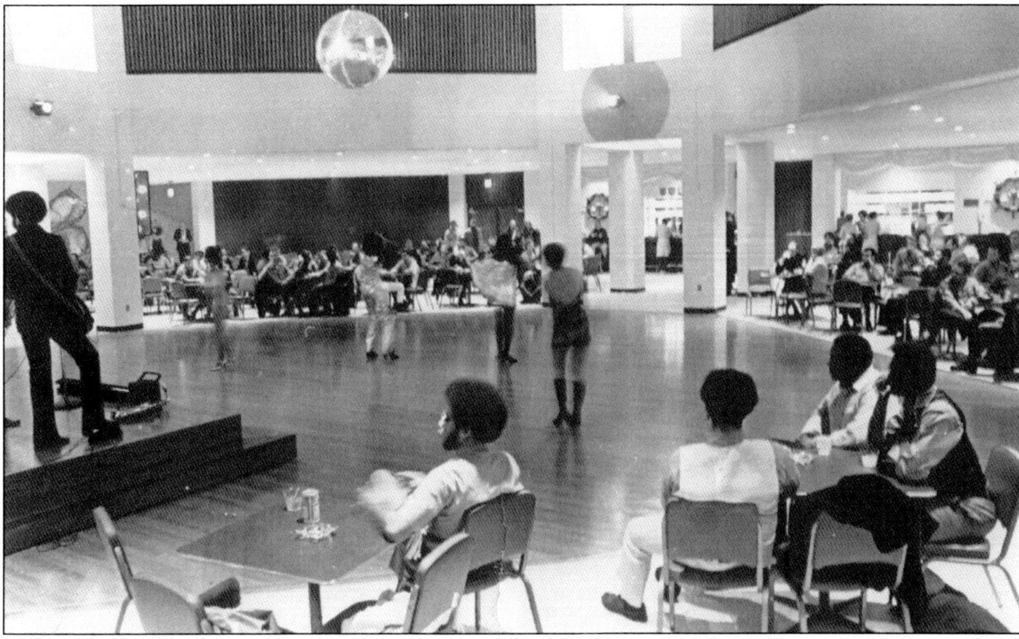

Morale has always been an important part of military life, and the Navy base at Millington is no exception. Pictured above in 1971, go-go dancers and disco music entertain the troops at the Seven Seas Enlisted Men's Club through the cold gloom of winter. There was also a petty officers' club and noncommissioned officers' club. Personnel, their dependents, and guests had access to various outdoor activities, such as horseback riding, fishing at Eagle Lake, a golf course and clubhouse, a skeet range, and swimming. Indoor recreation included watching movies at the base theater, a roller-skating rink, a library, and an Olympic-size indoor swimming pool (below), which was also used for water survival training. Base housing areas included playgrounds for dependents. Adults and children alike enjoyed treats from the McDonald's restaurant on base.

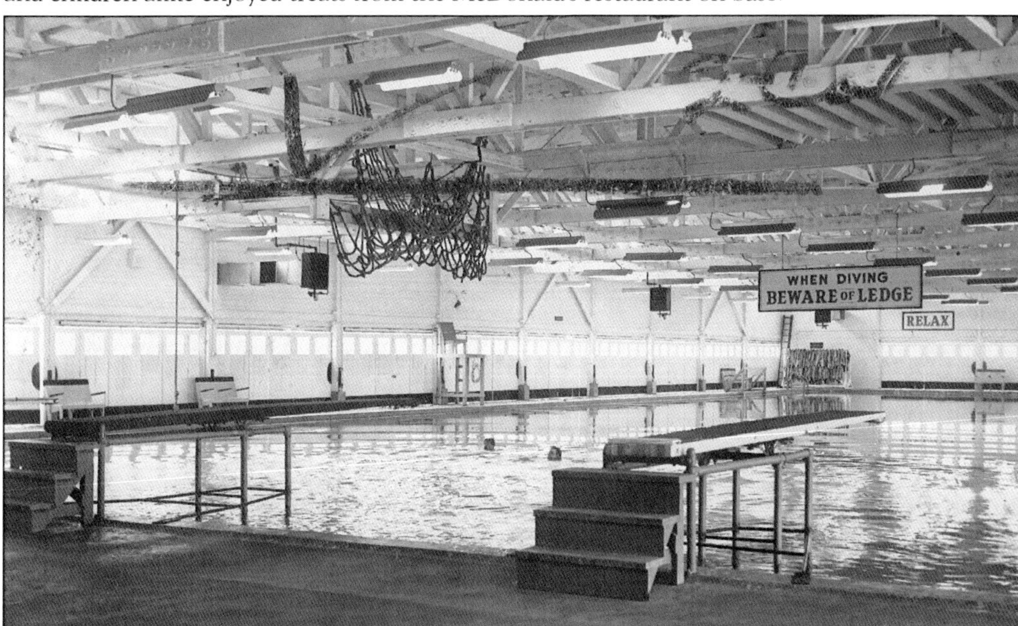

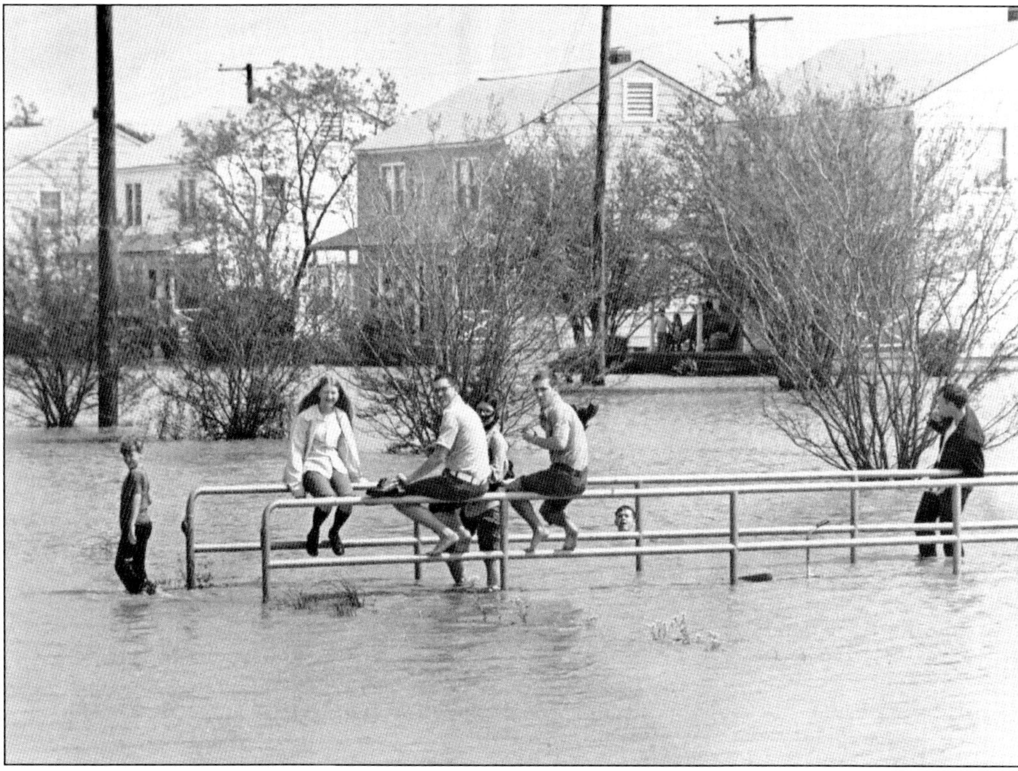

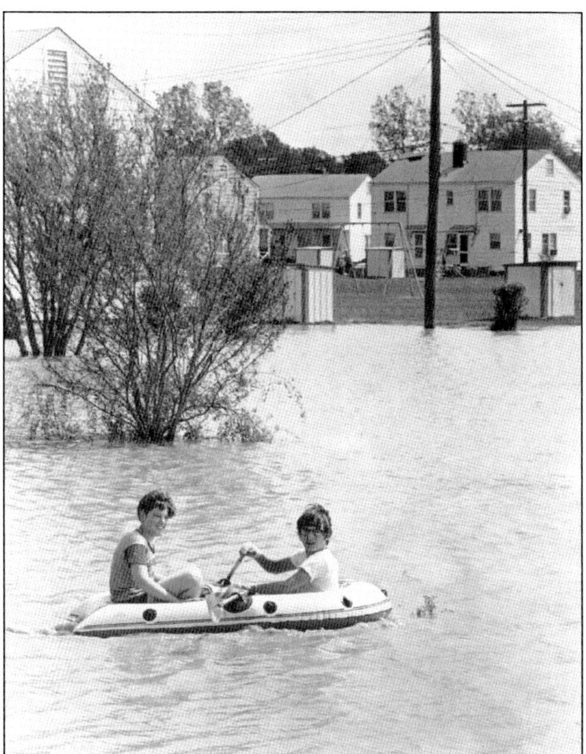

Millington has experienced the effects of several major floods, as the Mississippi River and its tributaries reached record levels, as well as several smaller incidents of flooding. The Great Flood of 1927 is perhaps the most famous, but other floods in 1937, 1973, and 2010 also wreaked havoc in Millington and on farms surrounding the town. Many farmers lost heavily in 1927 when some homes and farms literally caved in to the Mississippi River. W.R. Scott Jr. recalled helping his father move people from the river bottoms near Shelby Forest to a tent city set up on the Scotts' Herring Hill Road farm in 1937. These photographs from 1973 show how some residents near the naval base made the best of a bad situation while waiting for the high waters to recede.

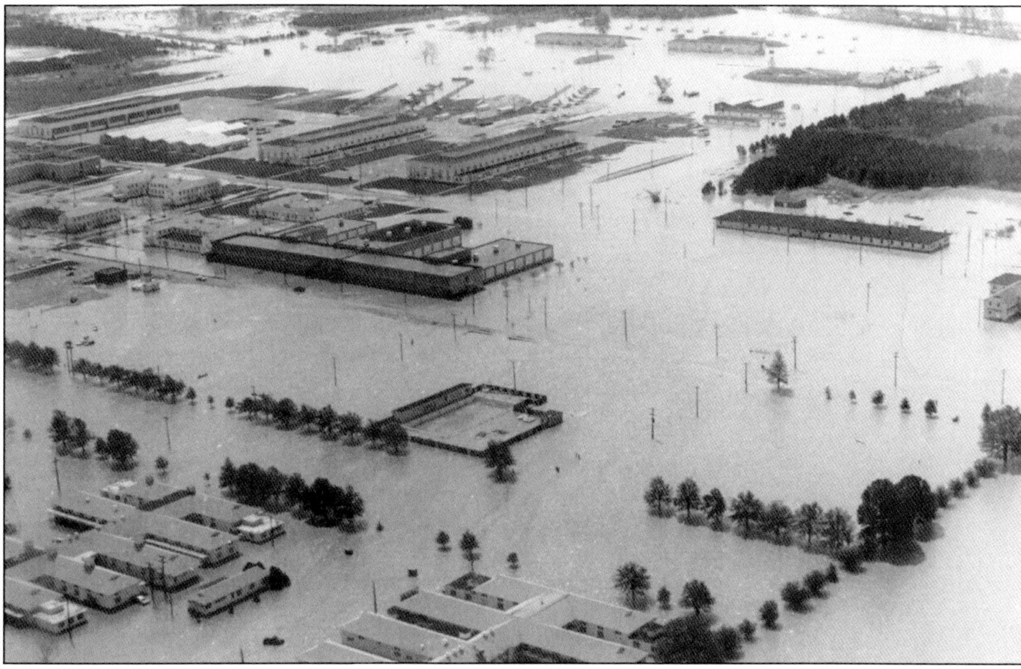

Buildings across the Navy base were inundated by floodwaters, as shown in this aerial photograph taken on April 21, 1973. Although the rising water eventually did reach the tarmac, none of the planes, seen in the top center of this photograph, were damaged. Ironically, few sailors appear to have been on deck when this image was captured. Periodic flooding continues to be a problem in the area.

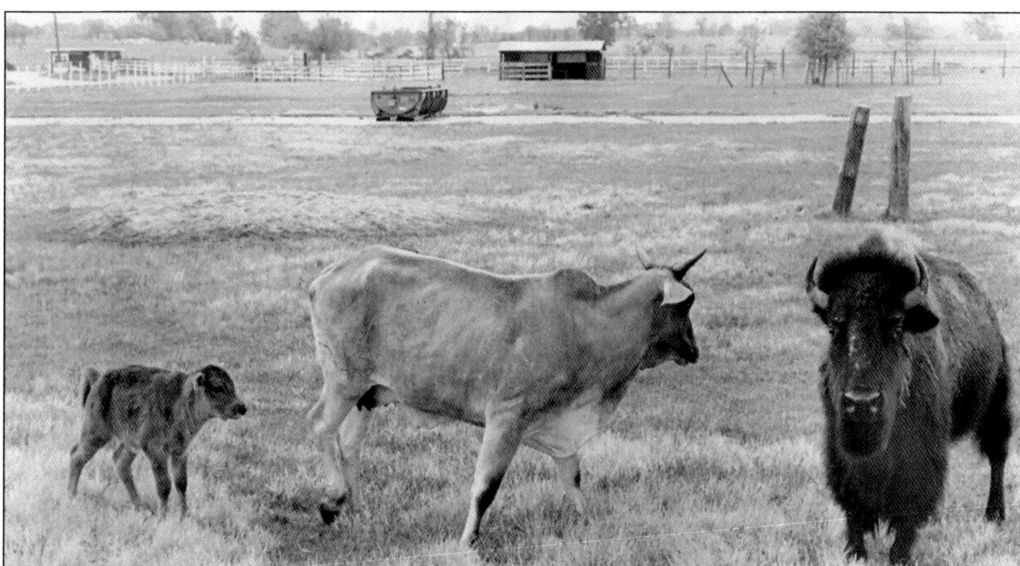

Many sailors arrived in Millington having never seen certain animals, so in 1973, the base began a petting zoo. Two Oklahoma buffalo, named Buffalo Bill Cody and Annie Oakley, were to be joined by elk, deer, sheep, and goats. The addition of cattle, however, had unforeseen consequences. The calf in the image, born in 1975 and named Little Eva, is the first known offspring of a Brahma cow and a buffalo.

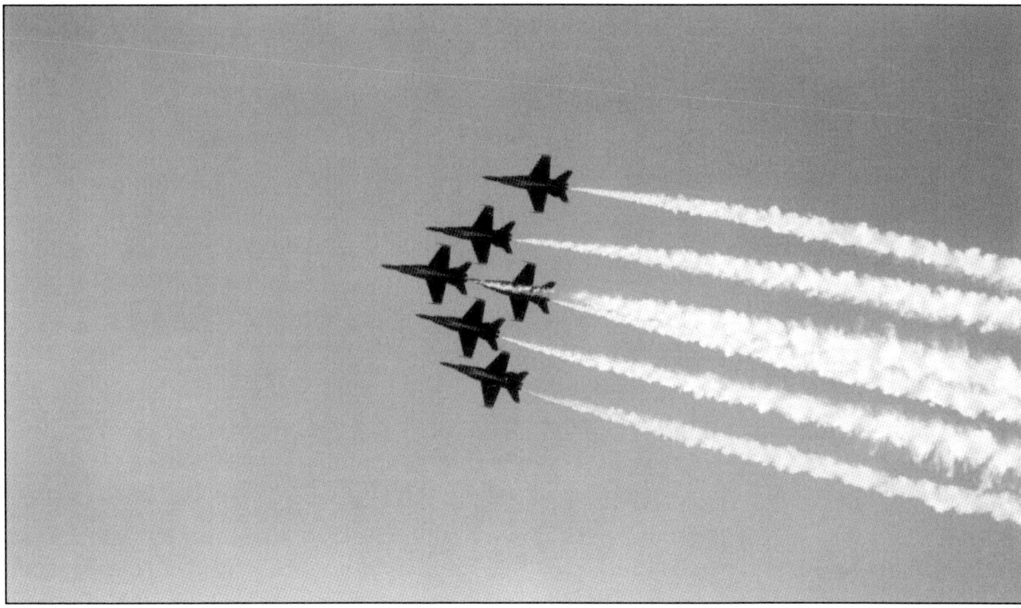

Air shows have long been a favorite attraction for Millington area residents, especially those featuring the Navy's Blue Angels. Pictured above, six jets fly in formation as jet trails slice through the sky. Undoubtedly, large crowds filled the grounds below to see all sorts of aerial feats. Similar shows date back to at least World War II, when wing-walker Phoebe Omlie delighted crowds at nearby Armstrong Field. The Blue Angels now have regular training exhibitions over Pensacola Beach, Florida. Below, area Boy Scout troops held an annual camp, Scout Base, at the naval air station. The boys and their leaders salute the flag as it is lowered so the sun will not set on it. (Courtesy *Millington Star*.)

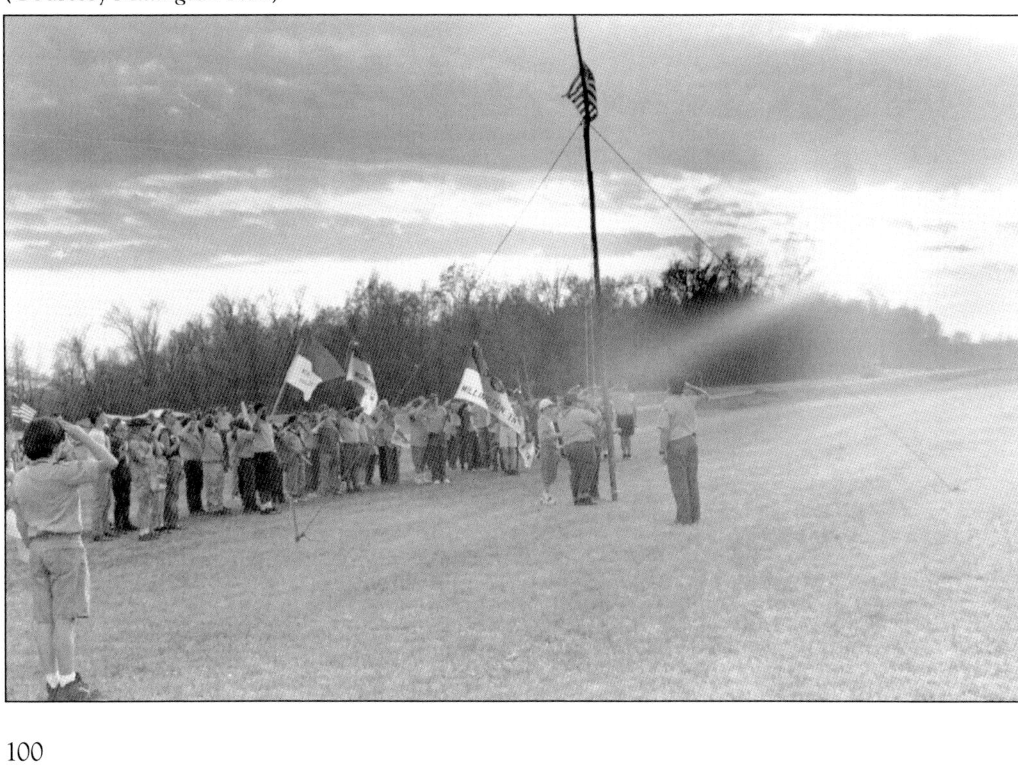

Four

Recreation and Entertainment

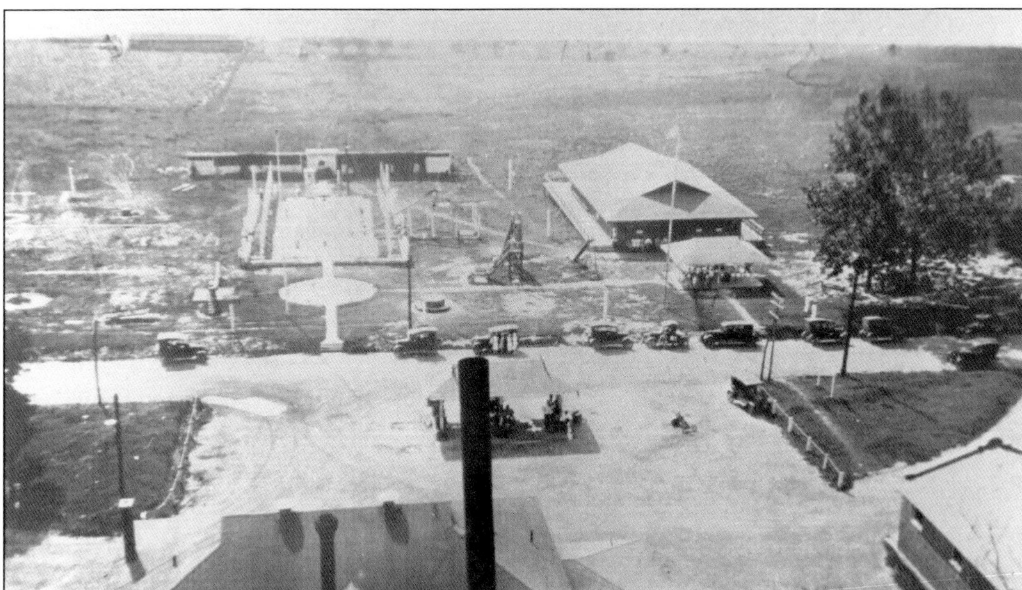

Hansonhurst, a resort near Bolton College, boasted a dance pavilion and public swimming pool (pictured), three stores, a bird sanctuary, icehouse with silent movie theater upstairs, playground, picnic area, and five electrified homes. Hansonhurst Creamery delivered butter, cream, and cheese by truck to Memphis stores and purchased cream from local residents. Founder Charles Hanson's utopian dream died in foreclosure in 1929, although some of the amenities remained available. (Courtesy MTC.)

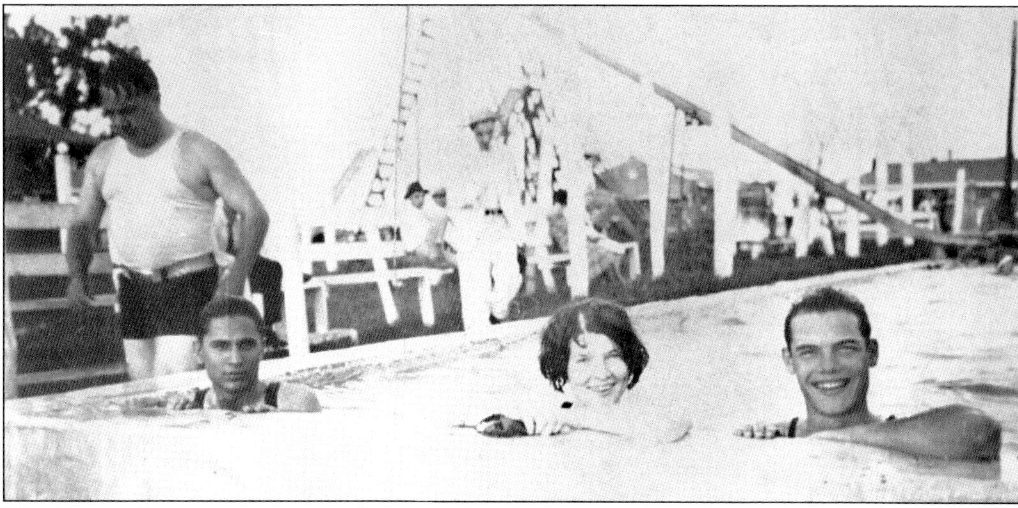

The Hansonhurst swimming pool was an enormous rectangle with a concrete wall that rose to about knee level and a tall slide, seen in the top center of the photograph above. A promenade led from the road to the pool and around it to the bathhouse. Admission to the pool was 35¢ or 50¢ with a rented swimsuit. An adjacent pavilion and dance hall featured live music by musicians like those pictured below and was the site of many sorority and fraternity socials and picnics the local girls liked to call "gypsy teas" because it sounded exotic. Alpha Tau Omega's Southwestern chapter, for instance, hosted a swimming party and picnic in May 1933, followed by a dance that evening. (Both, courtesy MTC.)

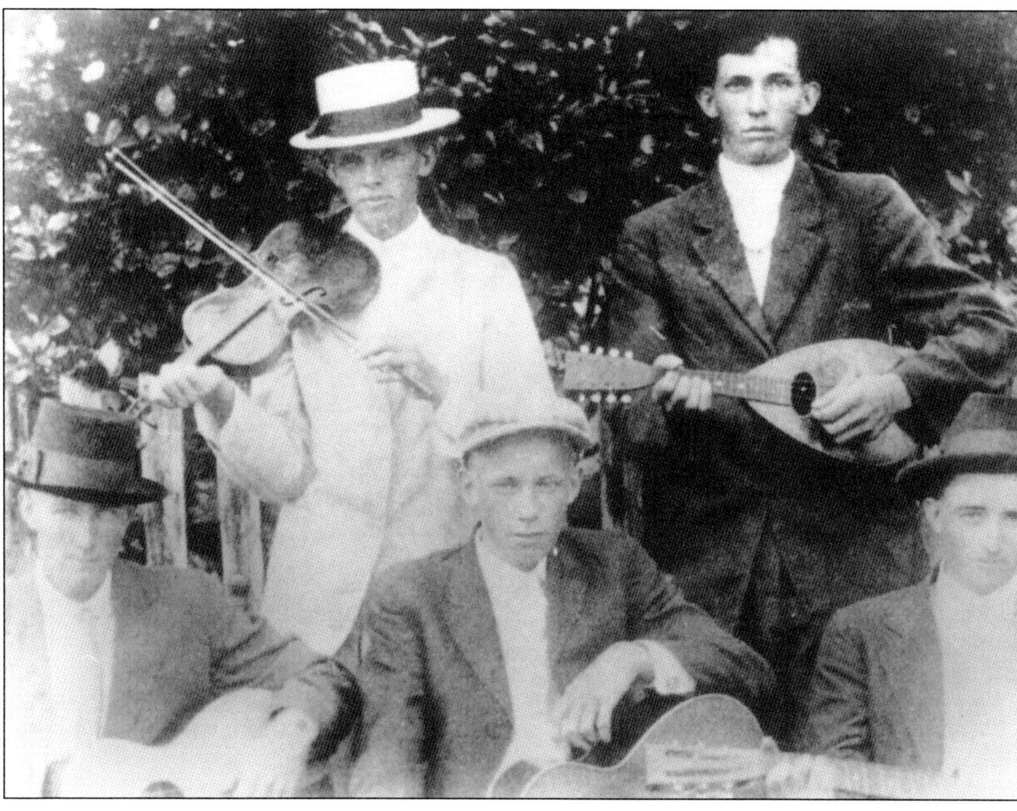

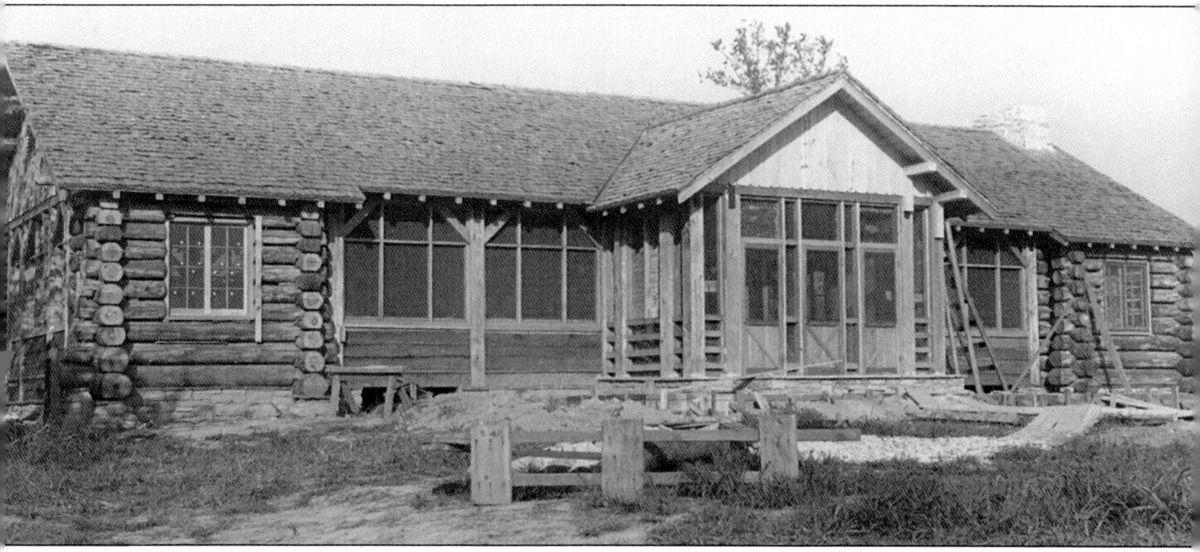

In 1934, the Shelby Forest Demonstration Project was still little more than a plan for a forestry and nature education and outdoor leisure area provided by Roosevelt's New Deal agenda, but it was not long before the park was open to the public. The first Shelby Forest Group Camp dining hall opened to the public in 1937. The building was utilized by Scout troops, church youth groups, students on school field trips, and for family reunions and other social gatherings. The camp served by this building was replaced by the Piersol Group Camp and a more modern dining facility. The Mississippi River Group Camp is now closed.

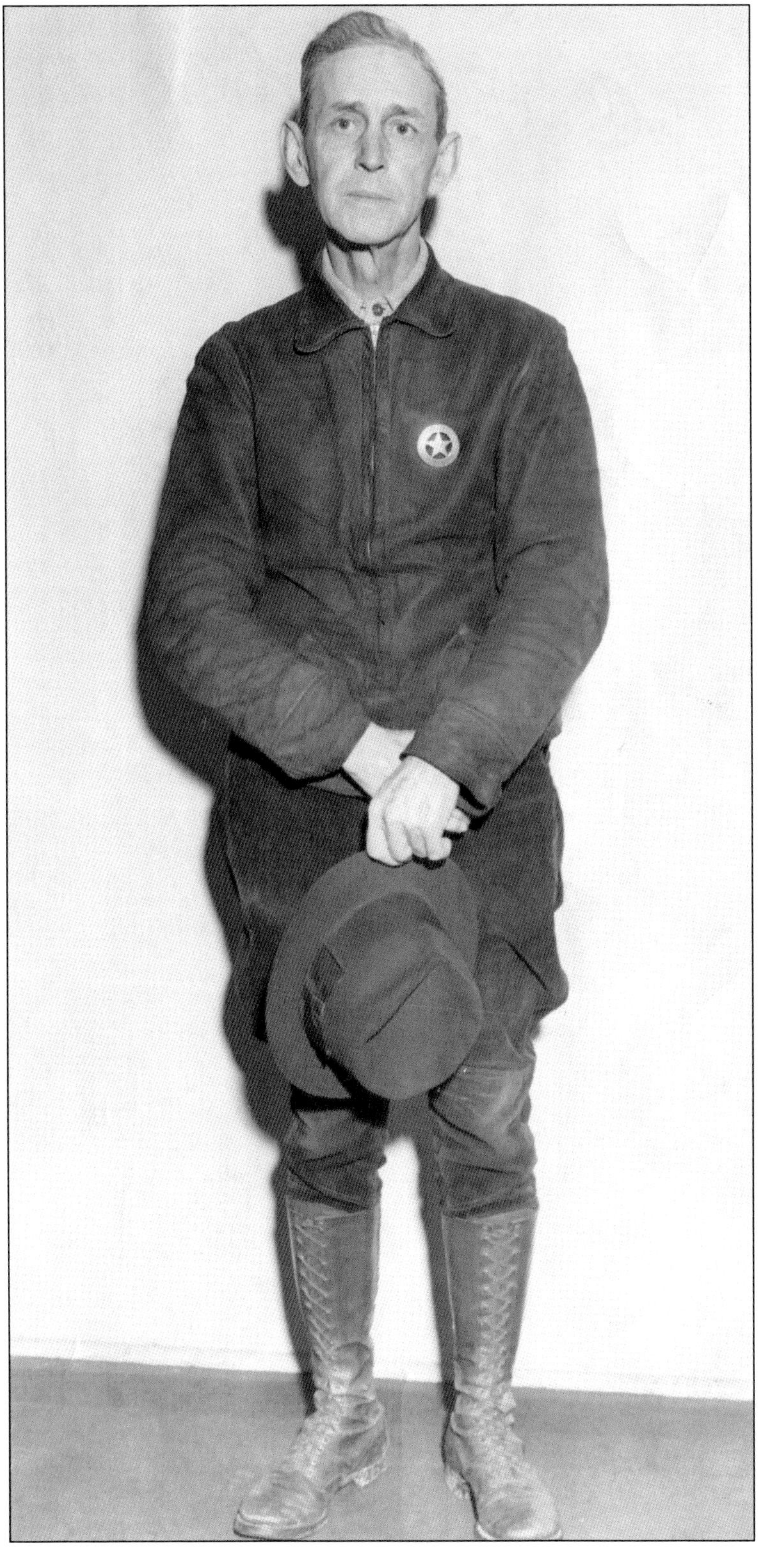

Pictured here in 1939, Shelby Forest's deputy US marshal, C.E. Courtney, declared the fox and the bobcat the "absolute rulers" of Shelby Forest. That year, six of the eight deer supplied by the federal government survived, and one fawn had been born. An even rarer sight was the otters. Other species in the park included wild turkey, snipe, doves, quail, woodcocks, ducks and geese, snakes, rabbits, and squirrels. Courtney's chief complaint was that visitors to the park were in such a hurry, they failed to appreciate the diversity of wildlife within the forest.

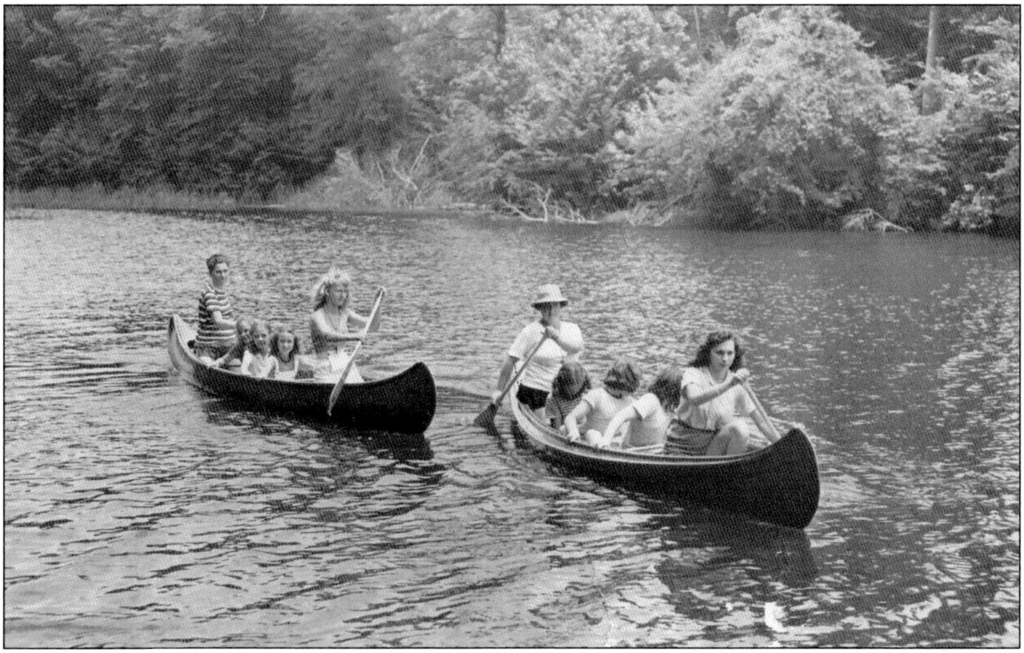

Canoeing on the park's several lakes offered visitors an opportunity to slow down and observe the park's natural beauty, like these Girl Scouts, who took advantage of canoe rental services one warm summer day in 1947. Piersol Lake is a natural lake from Shelby County's early days, but Poplar Tree Lake's 125-acre basin was dug in the 1930s. Grassy Lake in the state wildlife refuge area is no longer accessible.

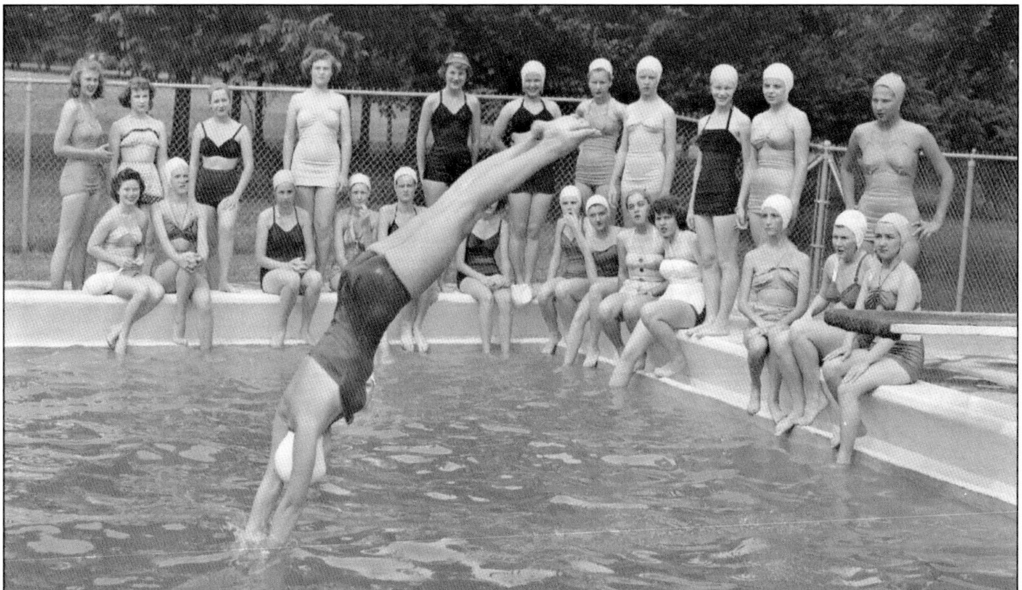

Pictured here in 1950, a group of young ladies critique the form of a diver at the Shelby Forest pool during a two-week camp. Although such camps allowed both genders, boys and girls were kept segregated both in the campground and in the pool. After the girls enjoyed swimming for an hour and a half, they relinquished the pool to the boys.

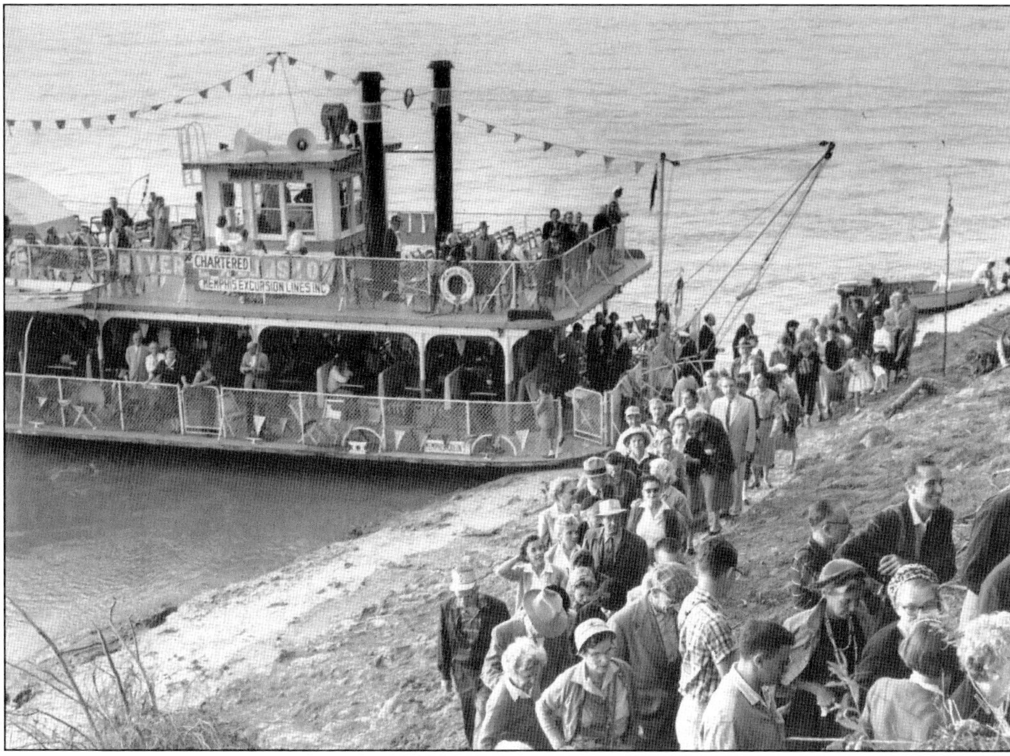

The *Memphis Queen* line of riverboats began offering Mississippi River sightseeing and educational excursions in 1954 under the management of Capt. Ed B. Langford. One popular stop for day cruises was the boat dock at Shelby Forest, where passengers could disembark and enjoy a day of picnicking, hiking, or bird watching before returning to Memphis late in the day. In October 1958, dozens of passengers forged upstream on the *Memphis Queen II*, a stern-wheeler first launched in 1955. The boat was frequently filled to its capacity of 256 passengers. By late 1955, construction of the Woodland Inn Restaurant (below) at Shelby Forest was nearing completion. Overlooking Poplar Tree Lake, the restaurant offered visitors standard picnic fare such as hot dogs and hamburgers. Other services included bicycle rentals, boat rentals, and bait sales.

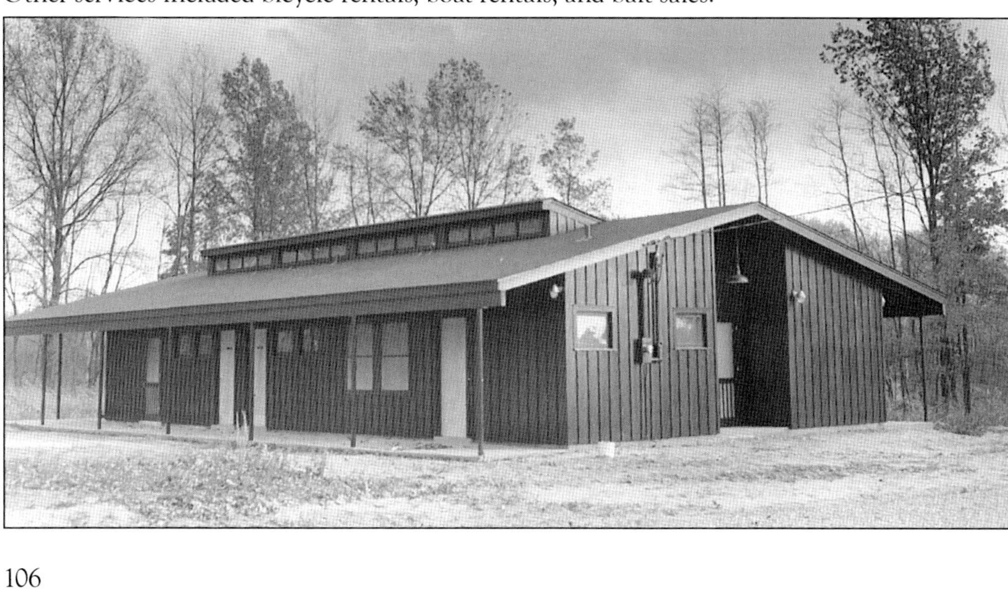

New cabins at Shelby Forest opened in 1962. The six cabins surrounding Poplar Tree Lake were designed for up to six people in each two-bedroom structure. Today, these cabins include kitchens, televisions with VCRs, linens, and, in winter months, wood for the fireplace. Nestled among mature trees, these cabins are typically booked well in advance.

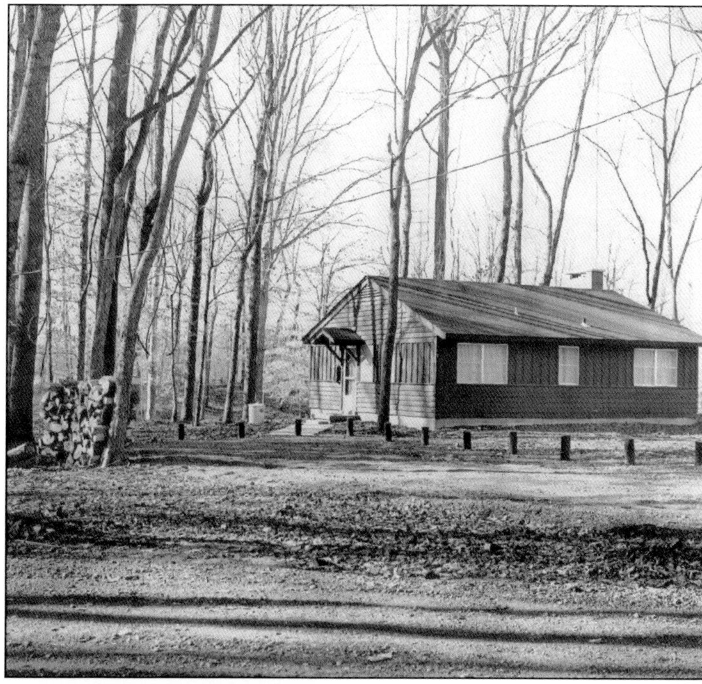

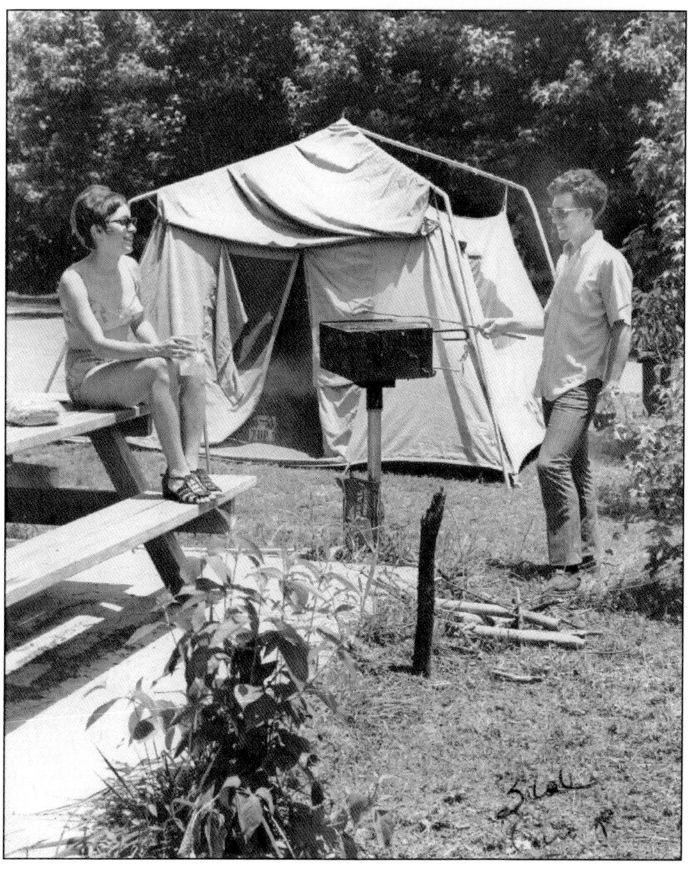

Camping at Shelby Forest draws Millington residents and people from around the world. Each of the park's 49 campsites includes a tent pad, picnic table, fire ring, grill, and water and electrical hookups. The campground also has a modern bathhouse with hot showers. Today, Shelby Forest rangers estimate annual attendance at 1.6 million visitors.

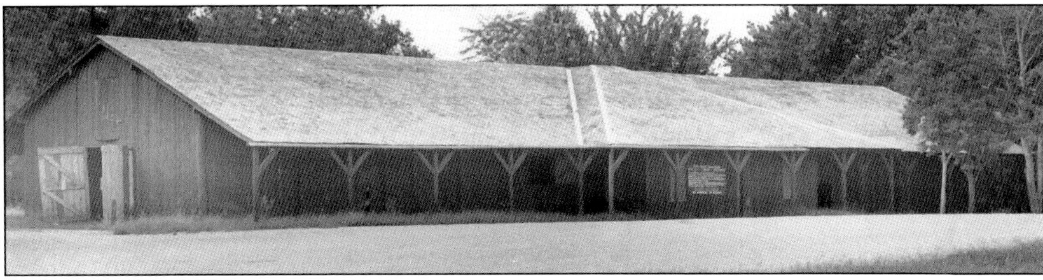

Situated near Grassy Lake, the horse barn seen above was completed in 1961. At least three saddle clubs have formed over the years, and all have enjoyed Shelby Forest's facilities. Pictured below during a 1961 exhibition at Embury Methodist Church, the Woodstock-Cuba Saddle Club has maintained a continuous presence at the arena on Woodstock-Cuba Road since 1969. The Shelby Forest Riding Club, established in 1966 by park superintendent Gordon Armour, Babe Howard, Layton and Gaston Edwards, Russell McFarland, Fred and Jimmy Hill, Ab Dix, and Wilson Sawyer, operated three years at the forest before moving to the Woodstock-Cuba arena. In 1971, the Benjestown Saddle Club lost two stallions, including a Tennessee walking horse, a mare, and a colt to a fire as well as a large barn and equipment. (Below, courtesy Doug Ammons, Shelby Forest General Store.)

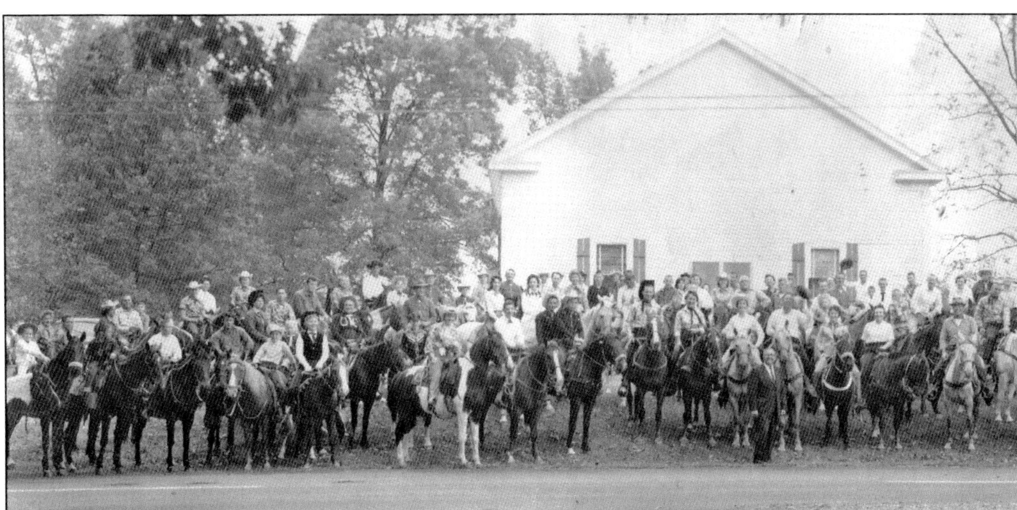

Many people have ridden Shelby Forest's eight miles of bridle paths since the first horses were introduced to the park in the 1960s. The horse trails wind through the trees, along the bluff, and around Poplar Tree Lake. Until recent years, rental facilities allowed individuals, friends and families, church and school groups, and other organizations to enjoy horseback riding in the park. Such outings often included picnics, fishing, or hiking before or after the ride. Horse rentals are no longer offered; however, area residents and visitors are welcome to bring their own horses to the forest trails, as always.

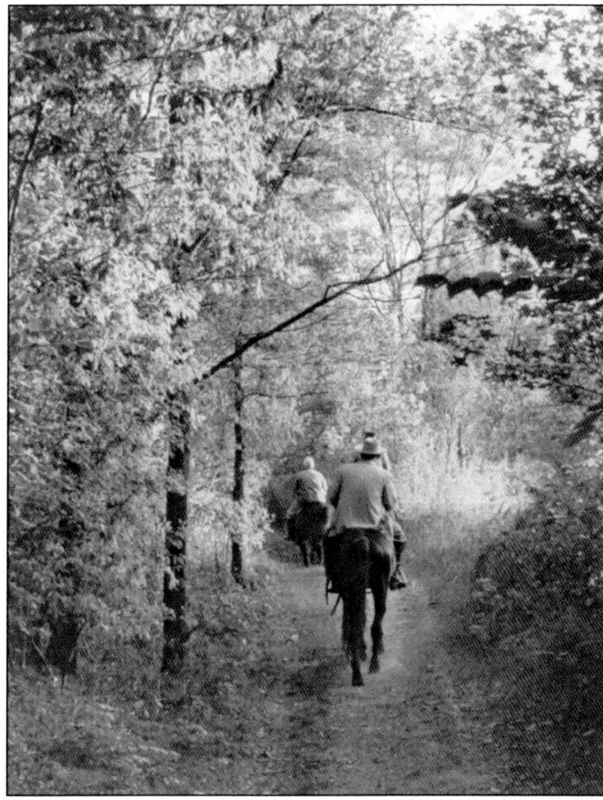

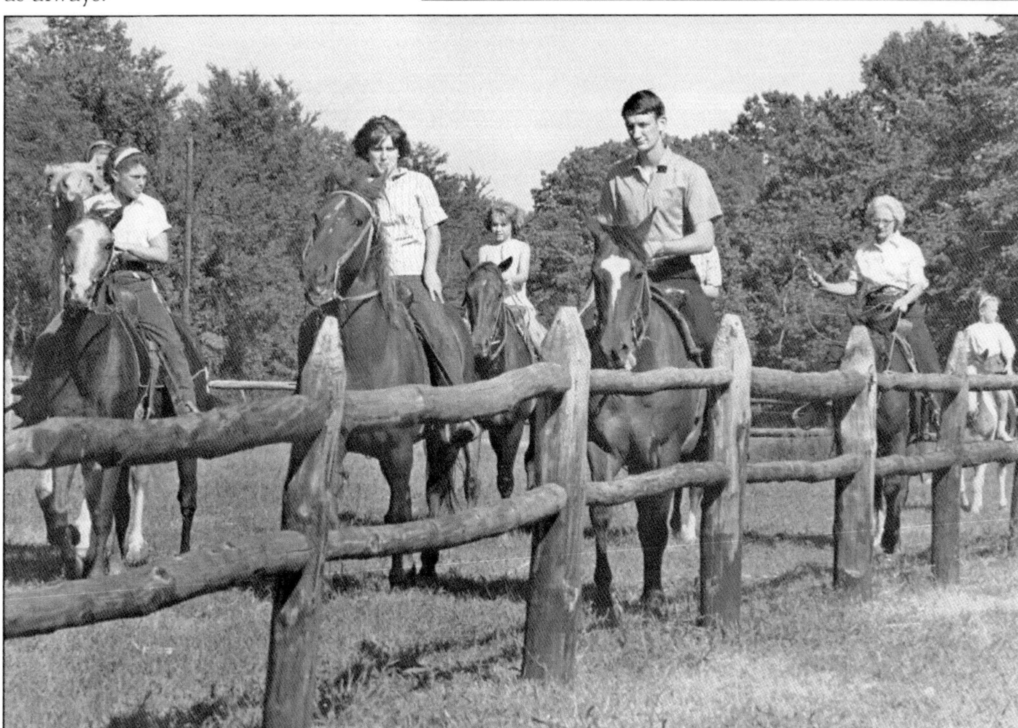

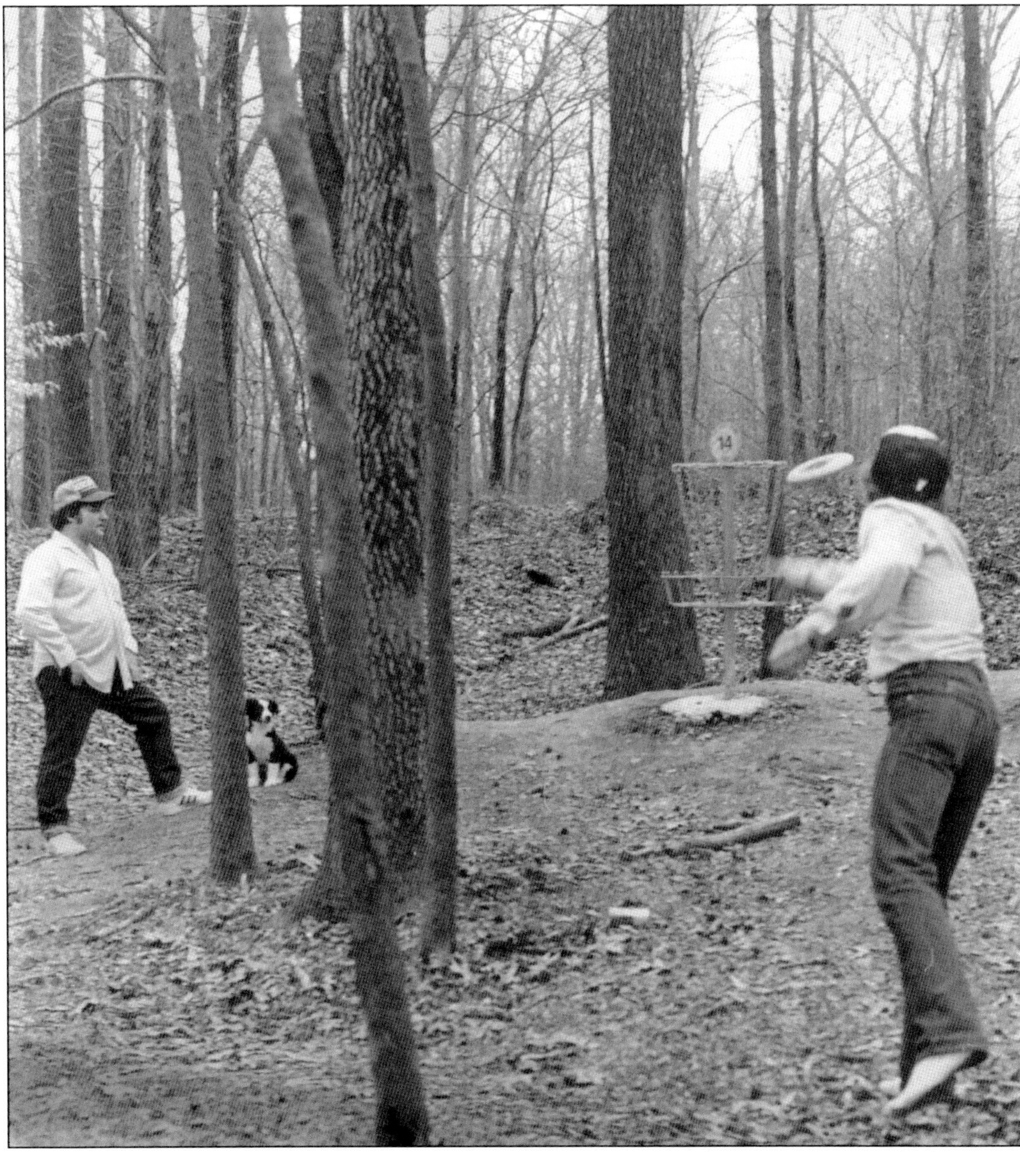

The Frisbee, or disc, golf course at Shelby Forest was added in the 1970s. Two 18-hole courses wind through the woods with goals tucked in behind trees, across gullies, or on the far side of hills, encouraging direct public interaction with the forest environment. Tournaments have drawn hundreds of visitors to the park on Saturday mornings, especially during the temperate spring and fall months.

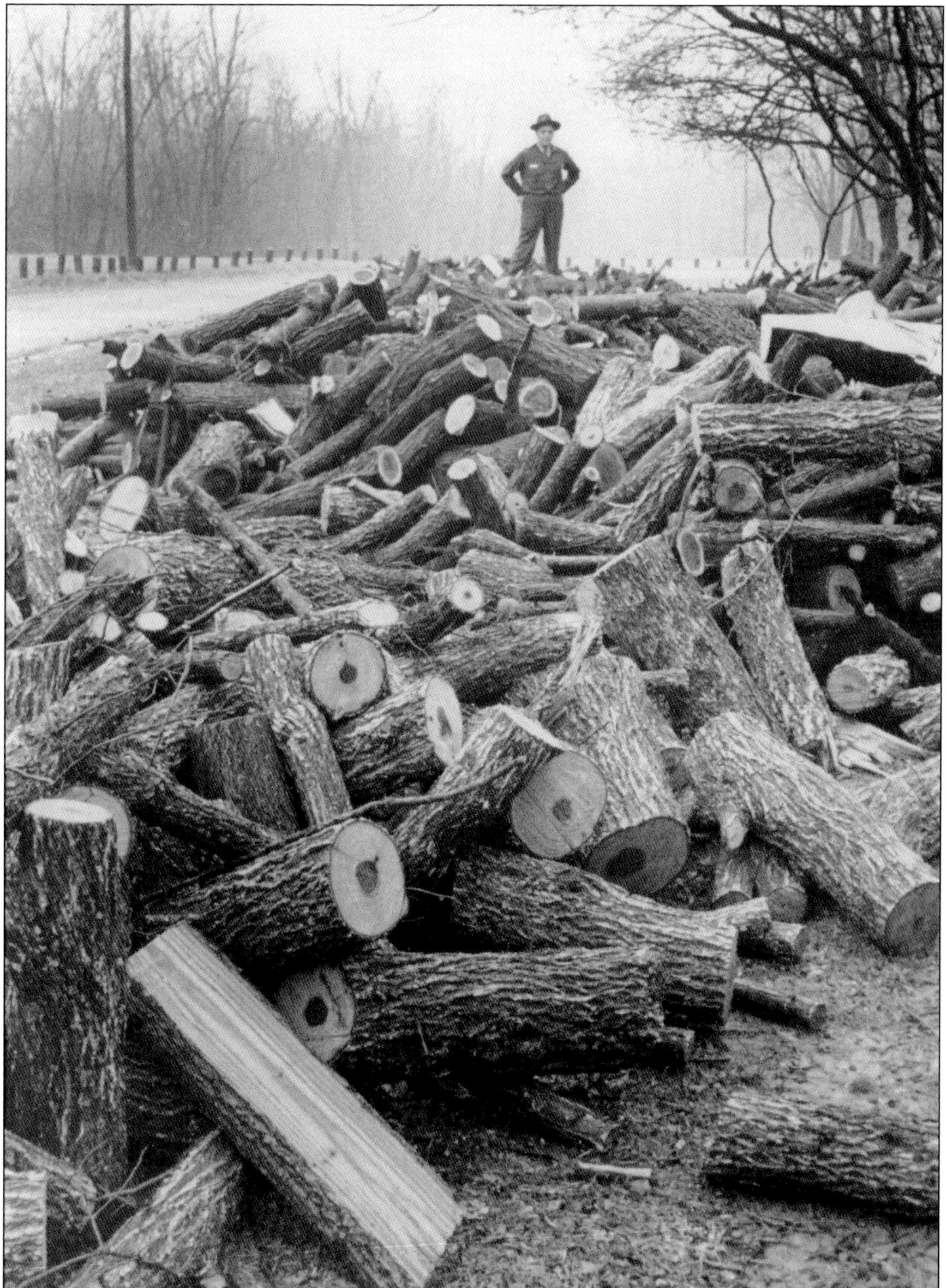

Park rangers are responsible for keeping both visitors and wildlife safe from one another and from natural disasters. Management of natural resources often includes removing dead trees to help prevent forest fires. Shelby Forest recycles much of this type of material by providing firewood for campers. Standing atop the firewood is park superintendent Neil C. Anderson.

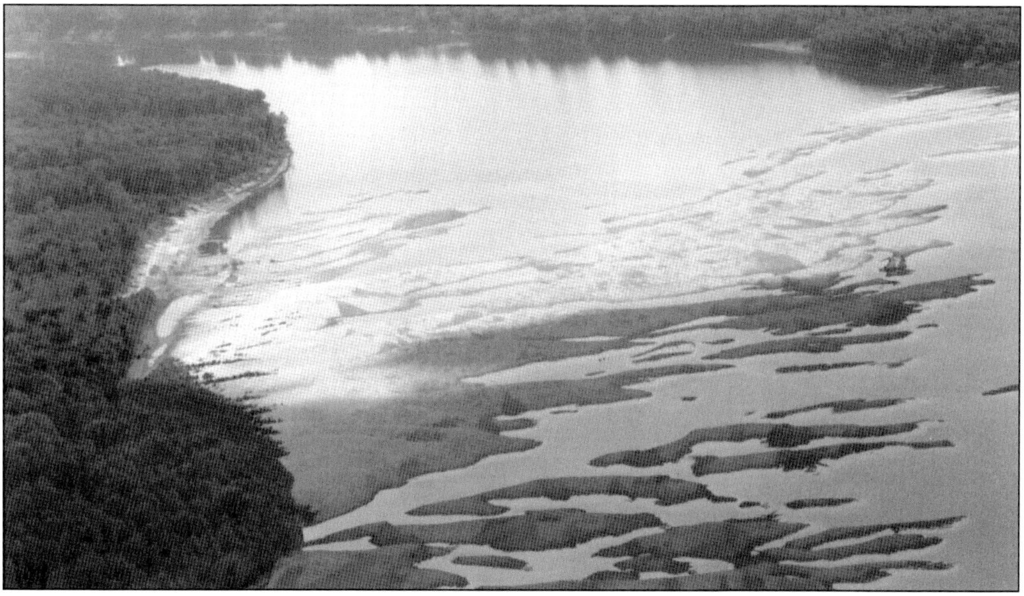

Although flooding is more common, occasional drought conditions occur, as pictured above in the mid-1980s. People from Millington, Memphis, and elsewhere flocked to the western edge of Shelby Forest to see the Mississippi River at a once-in-a-lifetime low. Those so inclined could literally walk across the river from Tennessee to Arkansas. (Courtesy *Millington Star*.)

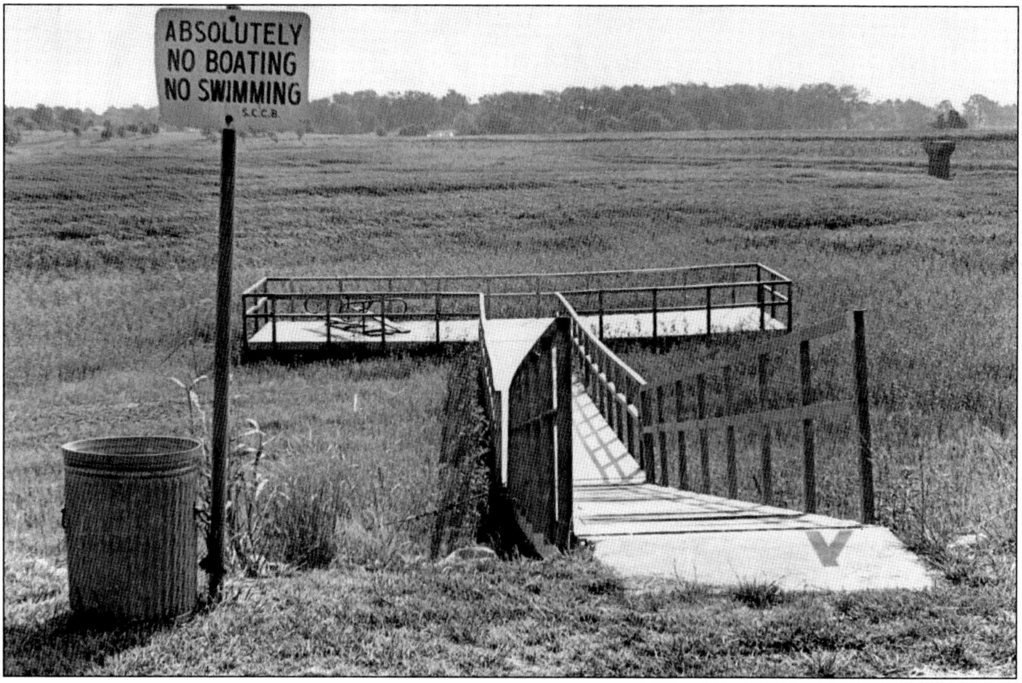

The Shelby County Conservation Board established Casper Creek Park in 1972. Renamed in honor of a former Memphis mayor, Edmund Orgill Park included a 45-acre man-made lake. Buildup of silt caused the county to drain the lake and remove 40,000 cubic feet of sediment, expanding the lake to its current 150-acre area. In 1976, a journalist quipped that with the lake dried up, Millington sailors truly were landlubbers.

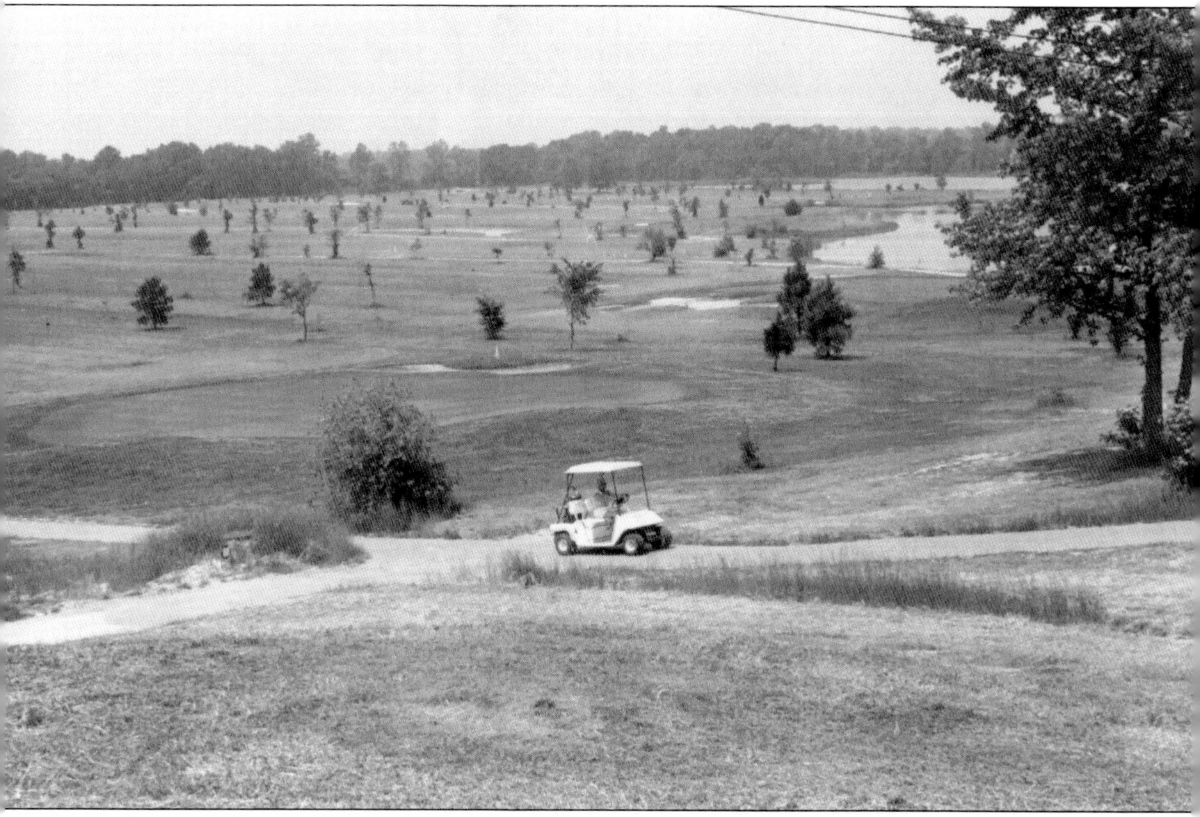

Dr. B.G. Mitchell opened the Woodstock Hills Country Club in 1978 on 300 acres he and his sisters inherited. A slow housing market interfered with efforts to establish a residential community surrounding the golf course pictured, and in 1990, the property was sold. Known for the next 19 years as Big Creek Golf Course, the property is now home to Justin Timberlake's Mirimichi Golf Course, one of four courses within seven miles of Millington. Edmund Orgill Park golf course is east of Millington, near Kerrville. The golf course occupies the western portion of Orgill Park, and the southeast corner was recently set apart for use by the West Tennessee Rugby Club. The Glen Eagle Golf Course opened in 1957 as a base amenity for military personnel but was opened to the general public following the base realignment in the 1990s. North of Millington at Drummonds, Forest Hills Golf Course provides Tipton County high school golf clubs a place to practice.

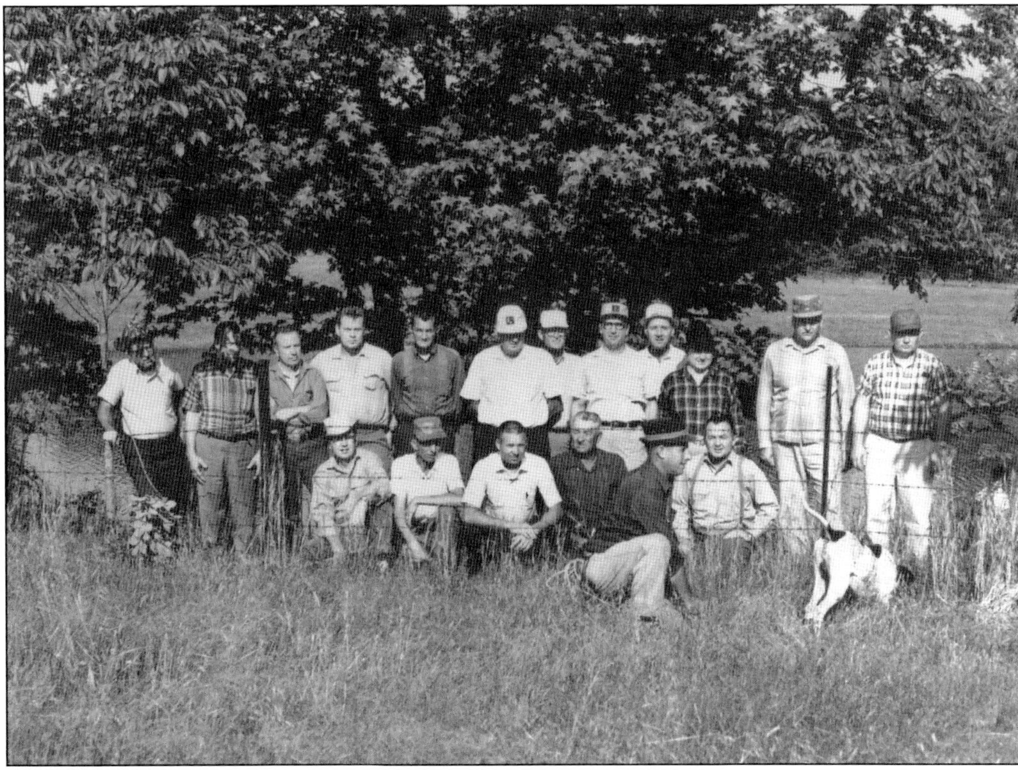

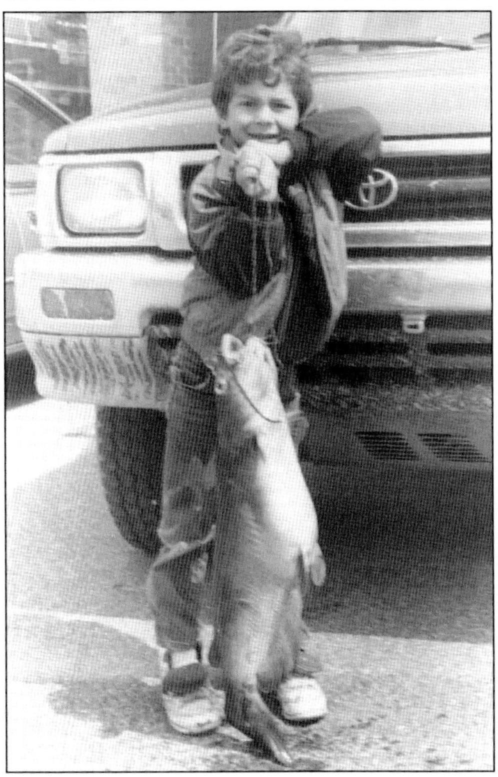

Outdoor activities have always been an important part of life for residents of rural areas, stemming in part from a deep connection with the land and the satisfaction of self-sufficiency. Babe Howard (above, kneeling third from right), wearing his trademark suspenders, was among the founders of the first quail-hunting club in Shelby County. Fishing is an age-old tradition shared by many rural families. At left, this youngster's catch was nearly as big as the boy, who struggles to hold it up for the camera. (Above, courtesy MTC; left, courtesy *Millington Star*.)

Hunting for some is more than sport; it is sometimes a necessity due to an overabundance of potentially dangerous wildlife. Verba Haynes, for example, watched first with amusement and then in horror as her neighbor, Evelyn Sigler, was attacked by a buck in her front yard near Shelby Forest. At right, a young man poses proudly with his rifle and his prey. Animals will also cause damage to crops, which undoubtedly prompted the capture of a wild hog near Herring Hill, as seen in the 1954 image below. A note on the back of the photograph identifies the hog's captors as, from left to right, Key, Walter, George, Lewis, and Nap. (Right, courtesy *Millington Star*; below, courtesy James Clifton and Jamie Scholik.)

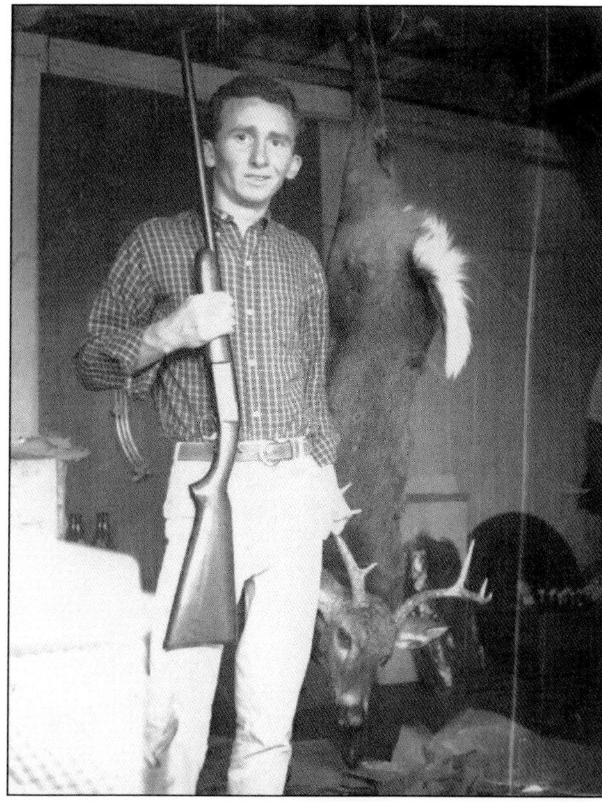

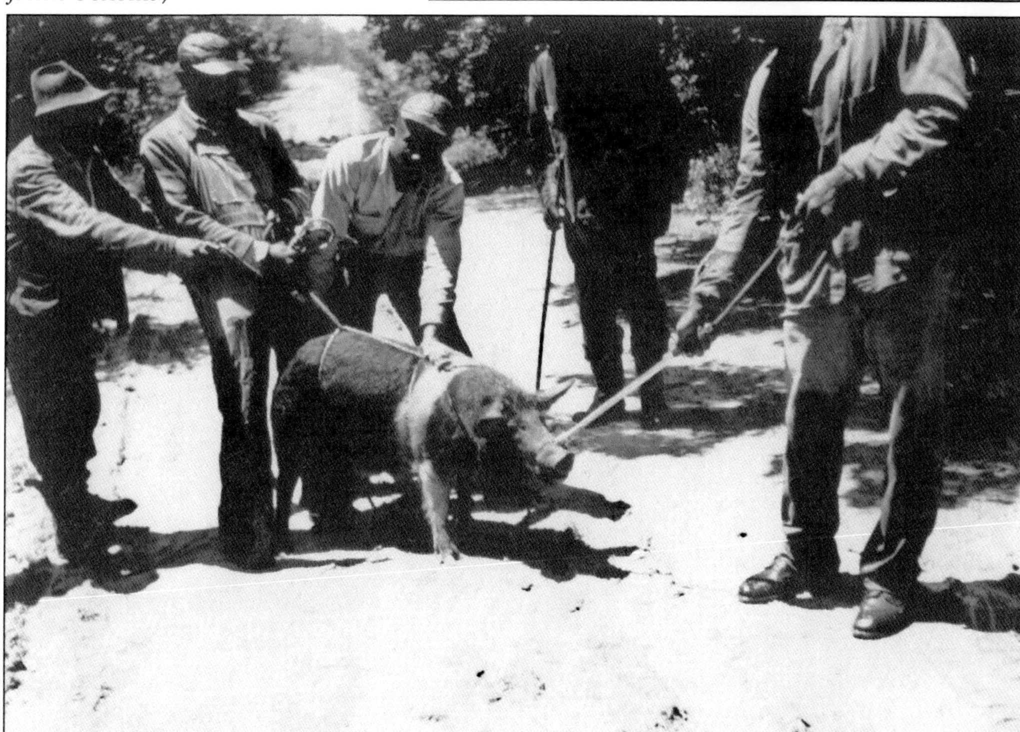

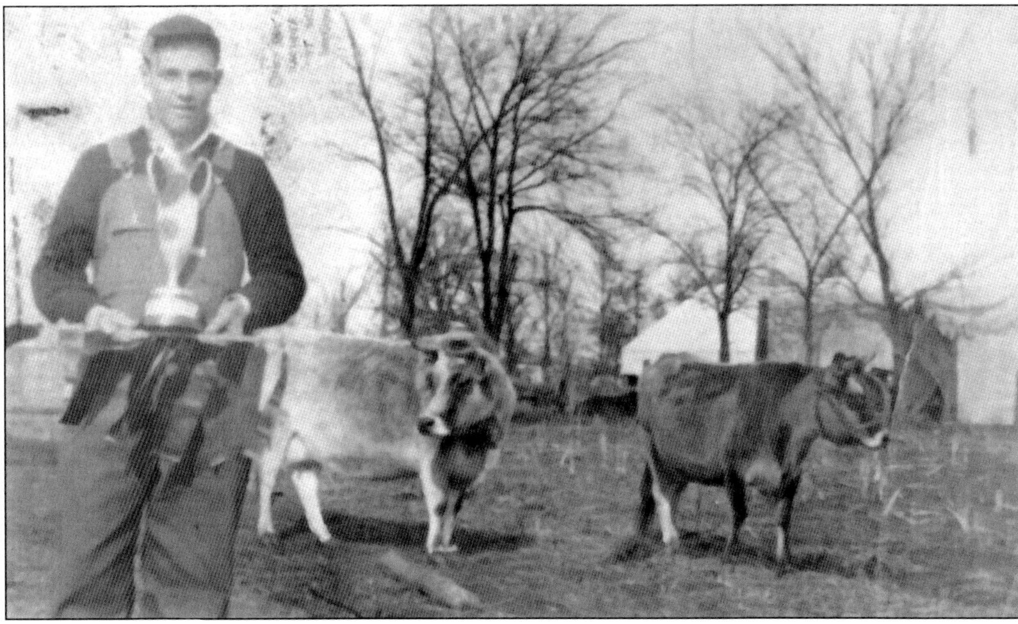

Other traditional activities center around agriculture, and many Millington area residents can boast of awards received at the Mid-South Fair, from the 4-H Club or from other events that judge the quality of homemade jams or preserves, quilts, horticulture, or livestock, as in the case of trophy winner L.M. Pace, above. An early example is that of Jack Scott, who won a new plow from the 4-H Club for being the county's "best boy plower" in 1935. More recent opportunities invite skeptical grins from onlookers, such as "chip-flipping" contests and chariot races at the annual Goat Days Festival at USA Stadium, seen below. (Above, courtesy MTC; below, courtesy *Millington Star.*)

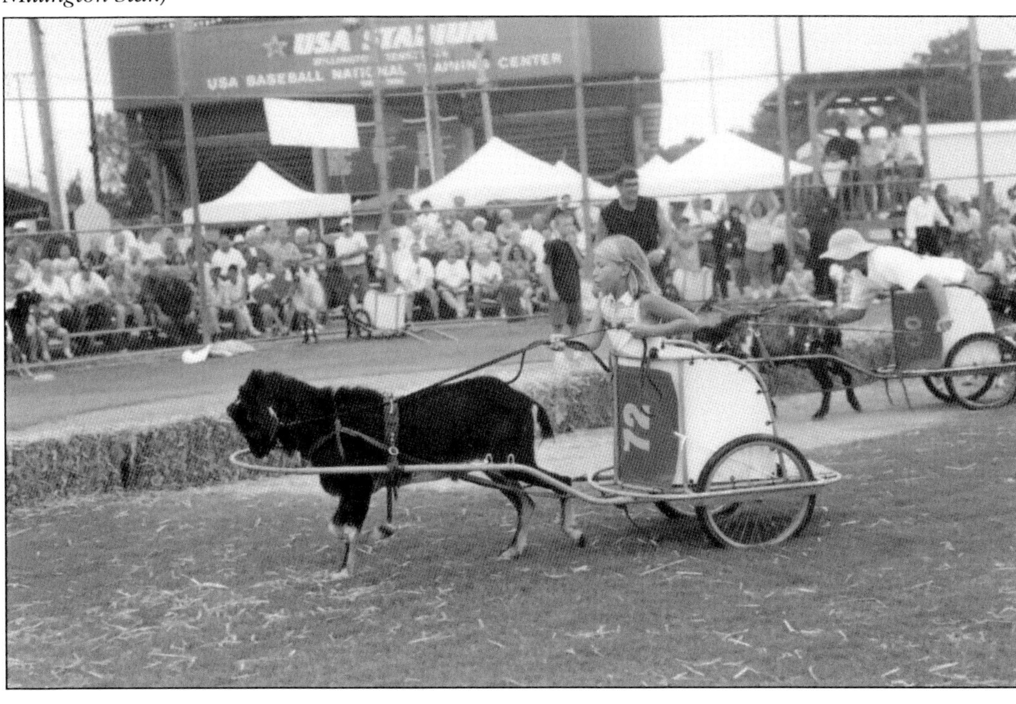

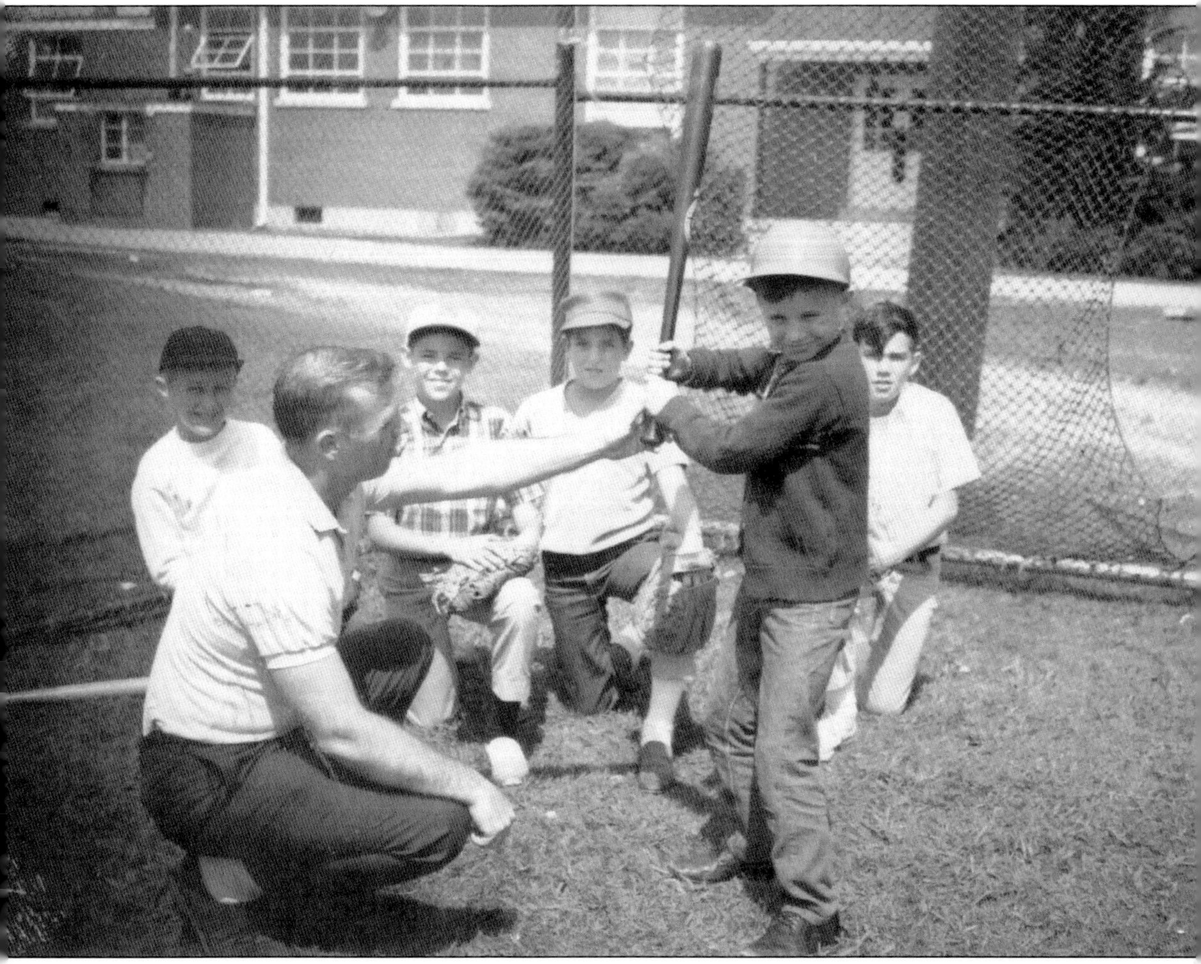

Baseball fields at local schools provided boys ample opportunity for practice of this locally favored sport. This nostalgic photograph could almost be a television still from the 1950s classic *Leave It to Beaver.* A community baseball diamond behind where the McDonald's now stands on Navy Road was another favorite place to play. (Courtesy *Millington Star.*)

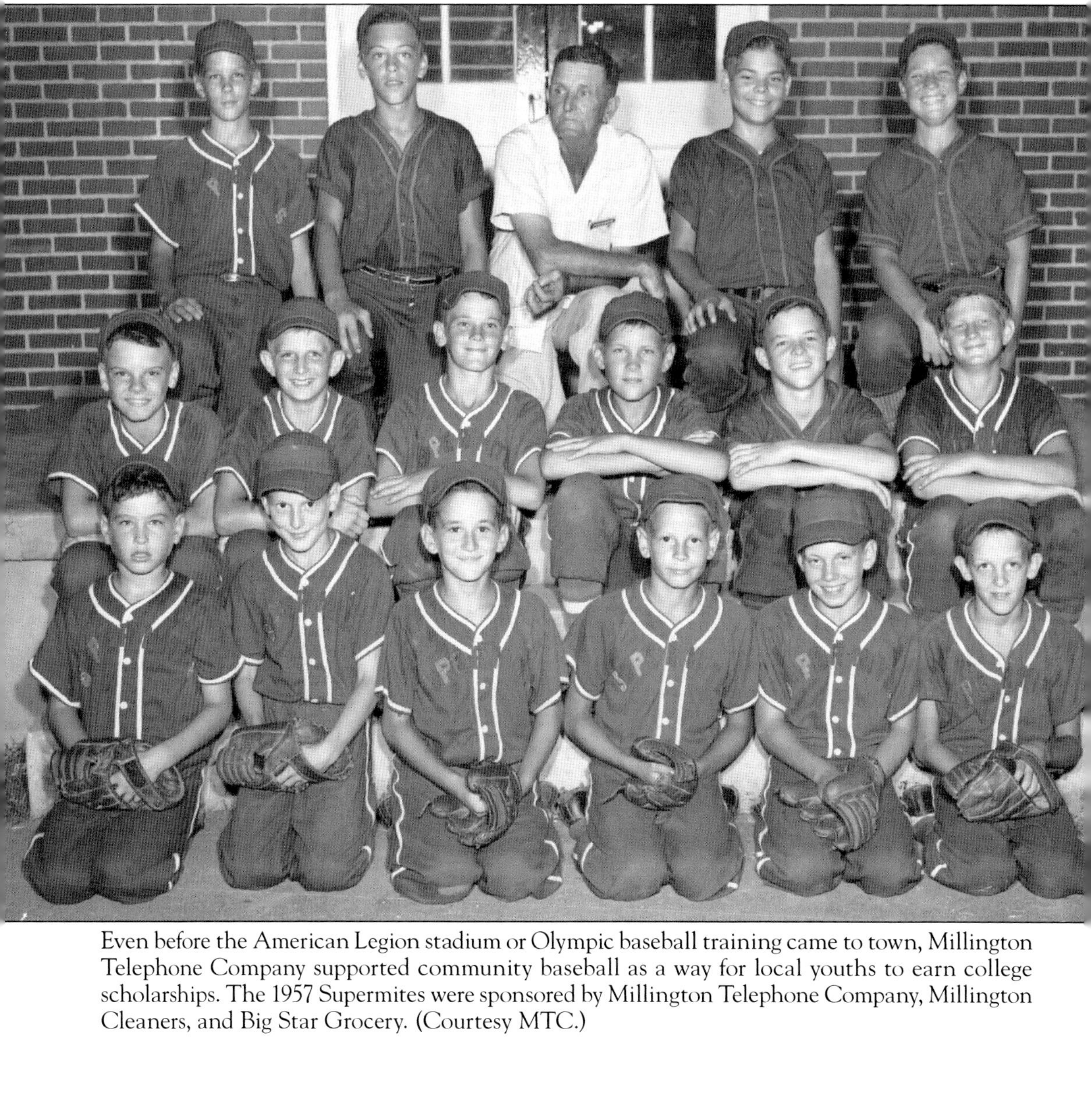

Even before the American Legion stadium or Olympic baseball training came to town, Millington Telephone Company supported community baseball as a way for local youths to earn college scholarships. The 1957 Supermites were sponsored by Millington Telephone Company, Millington Cleaners, and Big Star Grocery. (Courtesy MTC.)

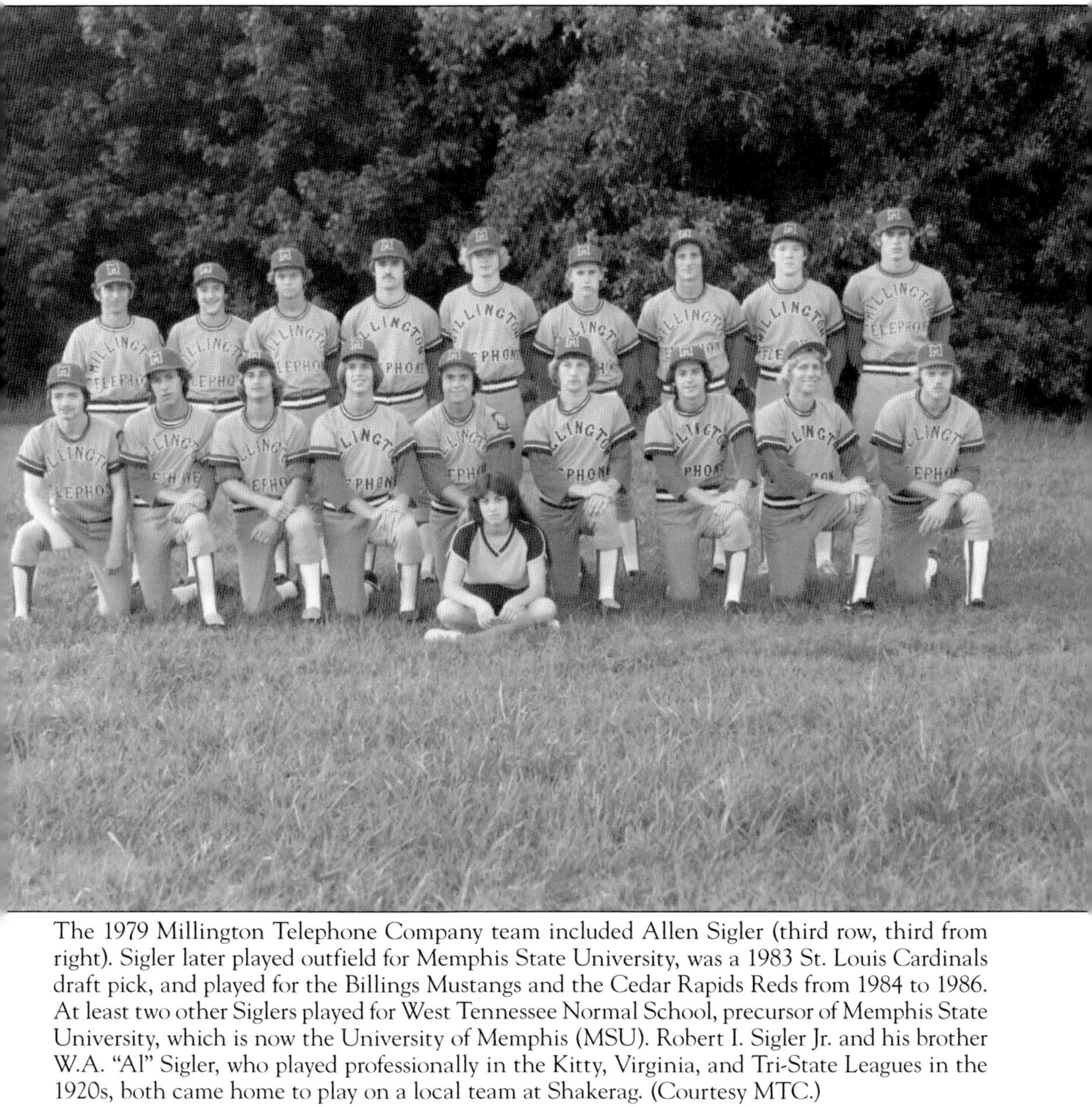

The 1979 Millington Telephone Company team included Allen Sigler (third row, third from right). Sigler later played outfield for Memphis State University, was a 1983 St. Louis Cardinals draft pick, and played for the Billings Mustangs and the Cedar Rapids Reds from 1984 to 1986. At least two other Siglers played for West Tennessee Normal School, precursor of Memphis State University, which is now the University of Memphis (MSU). Robert I. Sigler Jr. and his brother W.A. "Al" Sigler, who played professionally in the Kitty, Virginia, and Tri-State Leagues in the 1920s, both came home to play on a local team at Shakerag. (Courtesy MTC.)

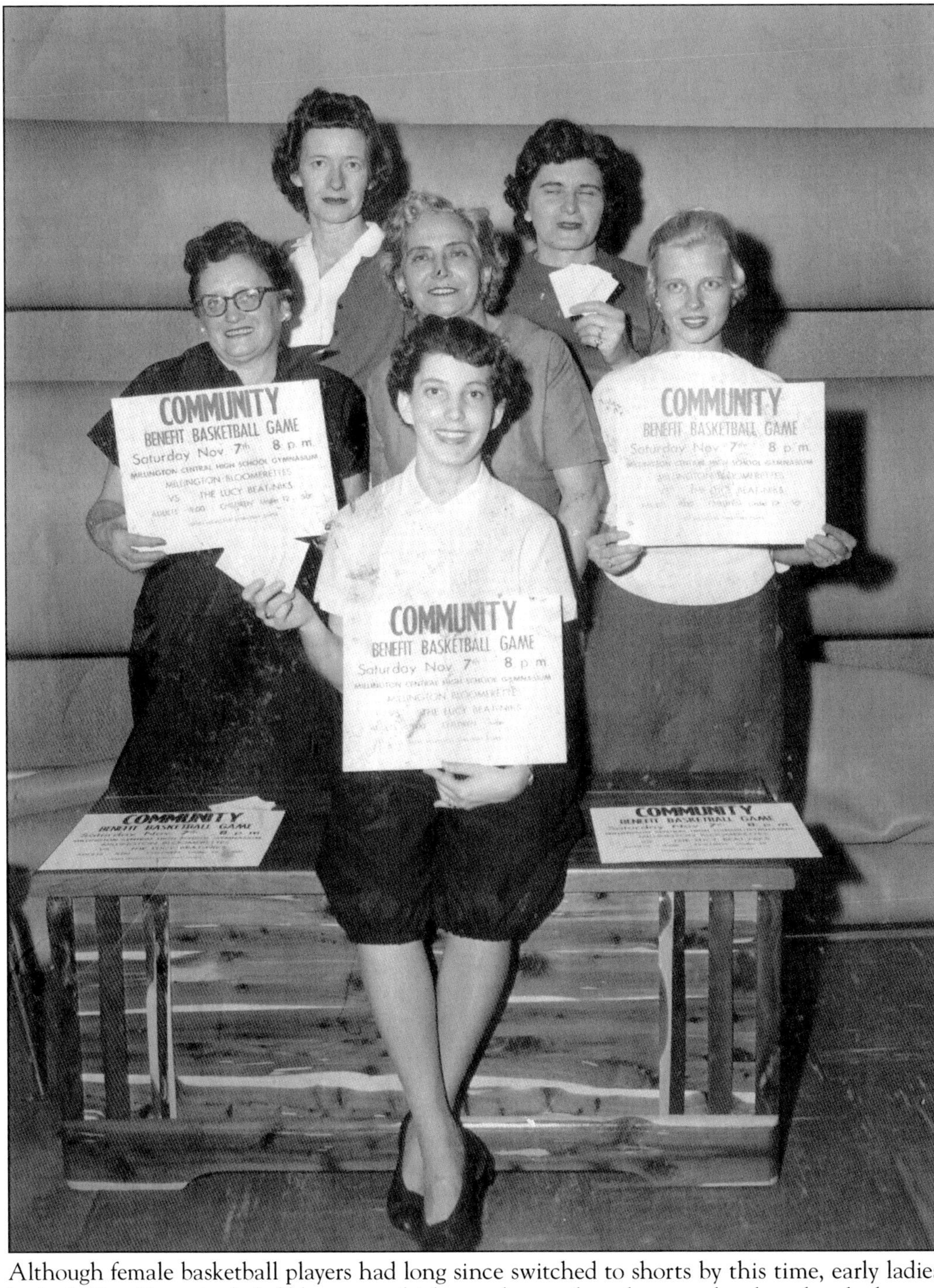

Although female basketball players had long since switched to shorts by this time, early ladies' basketball uniforms included bloomers, which were longer than shorts and gathered at the bottom of the leg, to protect a lady's dignity if she fell on the court. In 1959, a community basketball game to benefit the Millington Community Center pitted the Millington Bloomerettes against the Lucy Beat-Niks. Tickets were $1 for adults and half that for children under age 12. (Courtesy *Millington Star*.)

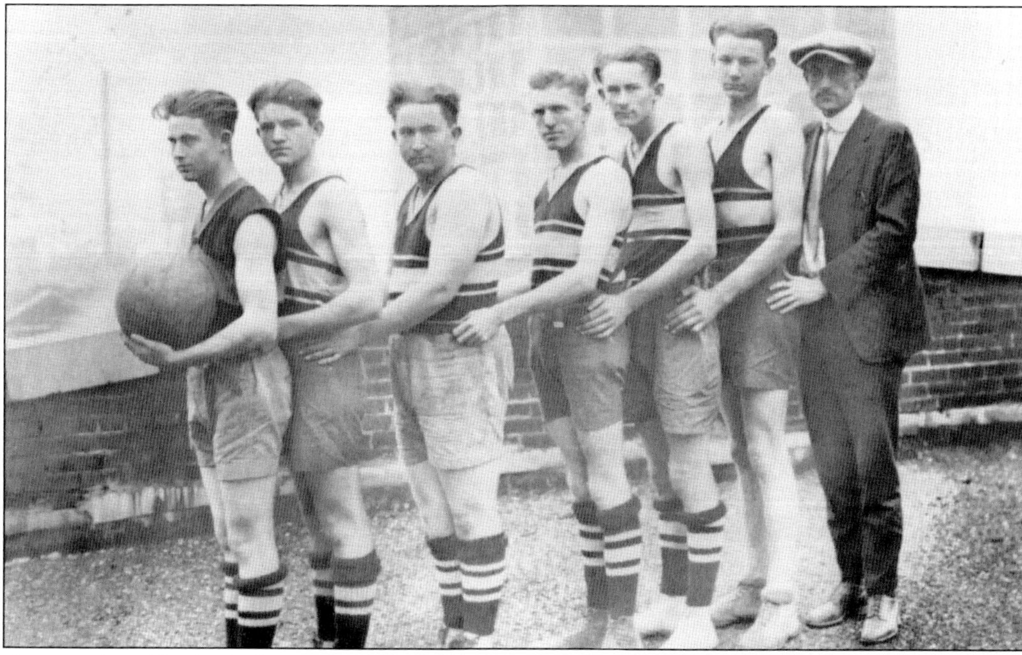

An early boys' basketball team travelled to games and was pictured on a rooftop (above), probably in Memphis. Below, the 1929–1930 girls' basketball team from Millington Central High School posed for a group photograph with their trophy. After 1934, Millington's teams were proud to host home games in their new gymnasium. (Both, courtesy MTC.)

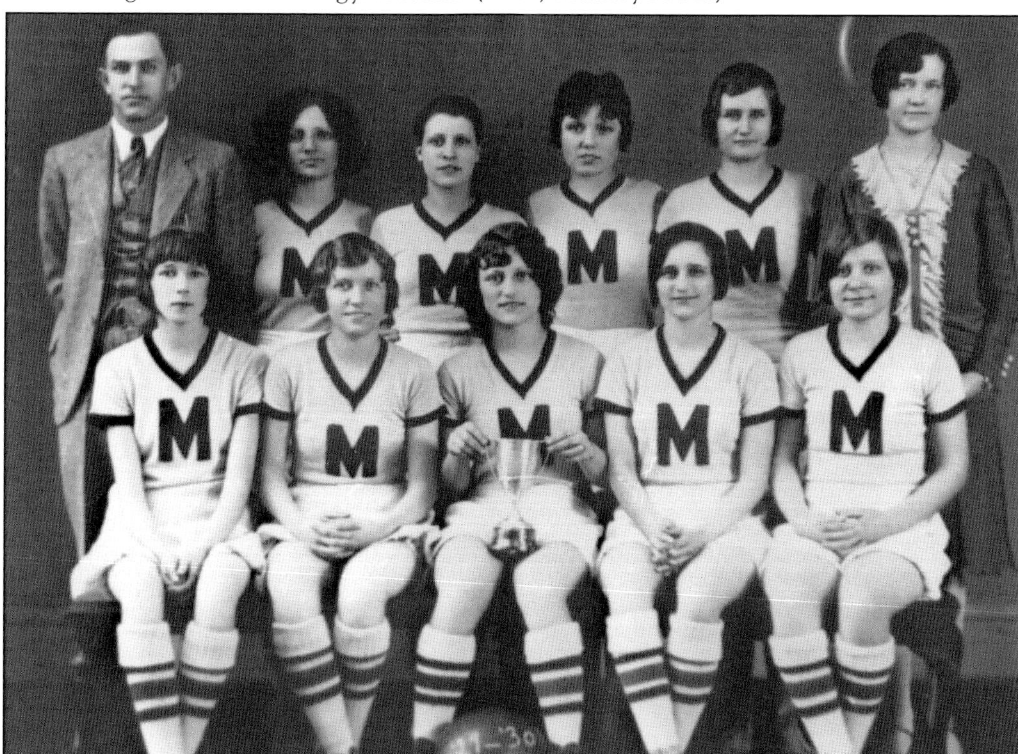

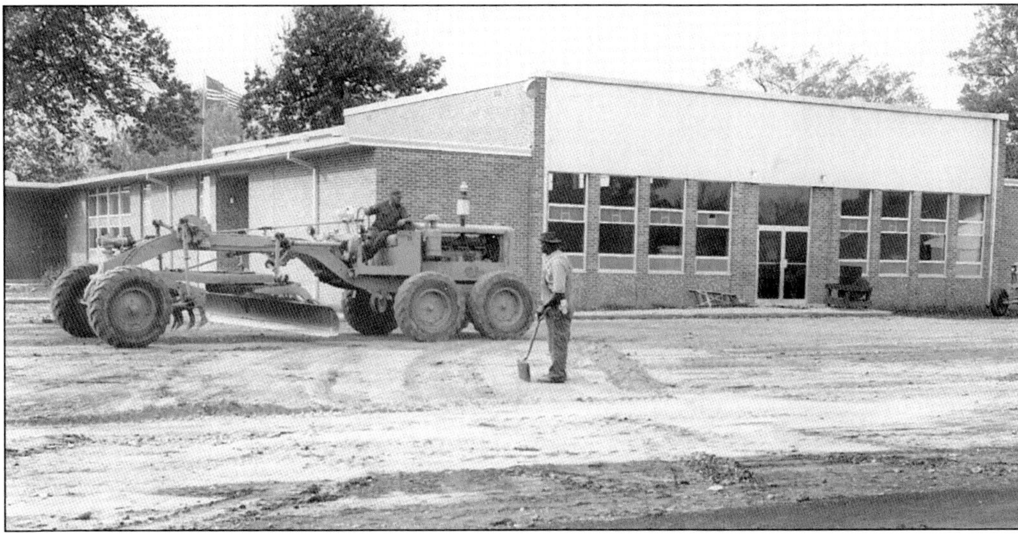

In November 1961, an asphalt parking lot was laid at the community center, in part to accommodate the increased traffic expected as the Millington branch library made its temporary home here while the new library was under construction. The community center opened in 1960 and was named for Millington's first full-time mayor, Charles W. Baker, who was then chairman of the Shelby County Quarterly Court. Oak Park, below, opened next to Buford Trailer Park in 1953. Millington's older children liked to ride their bicycles to Oak Park and younger children enjoyed the wading pool and playground.

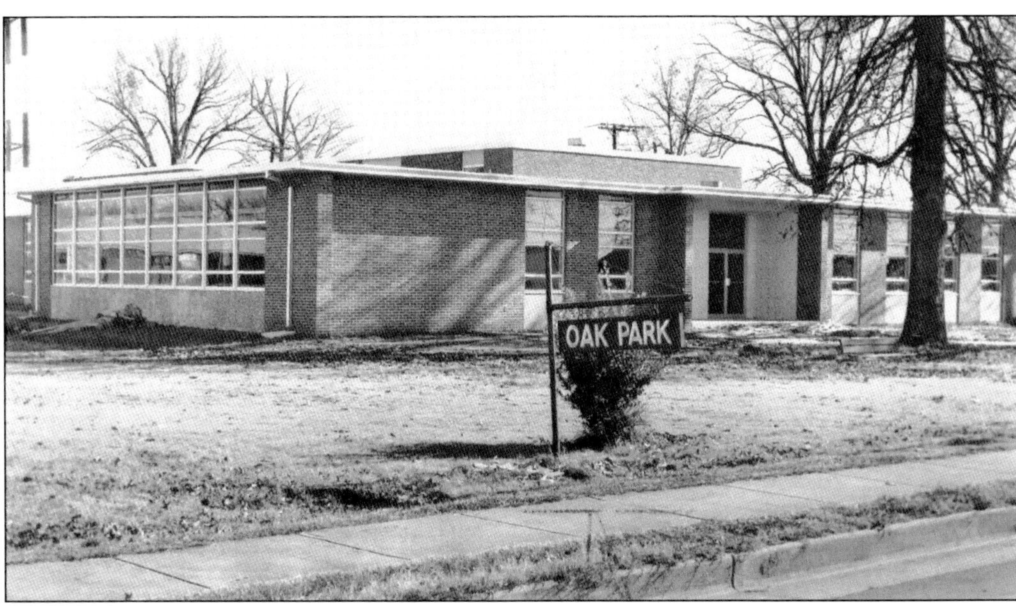

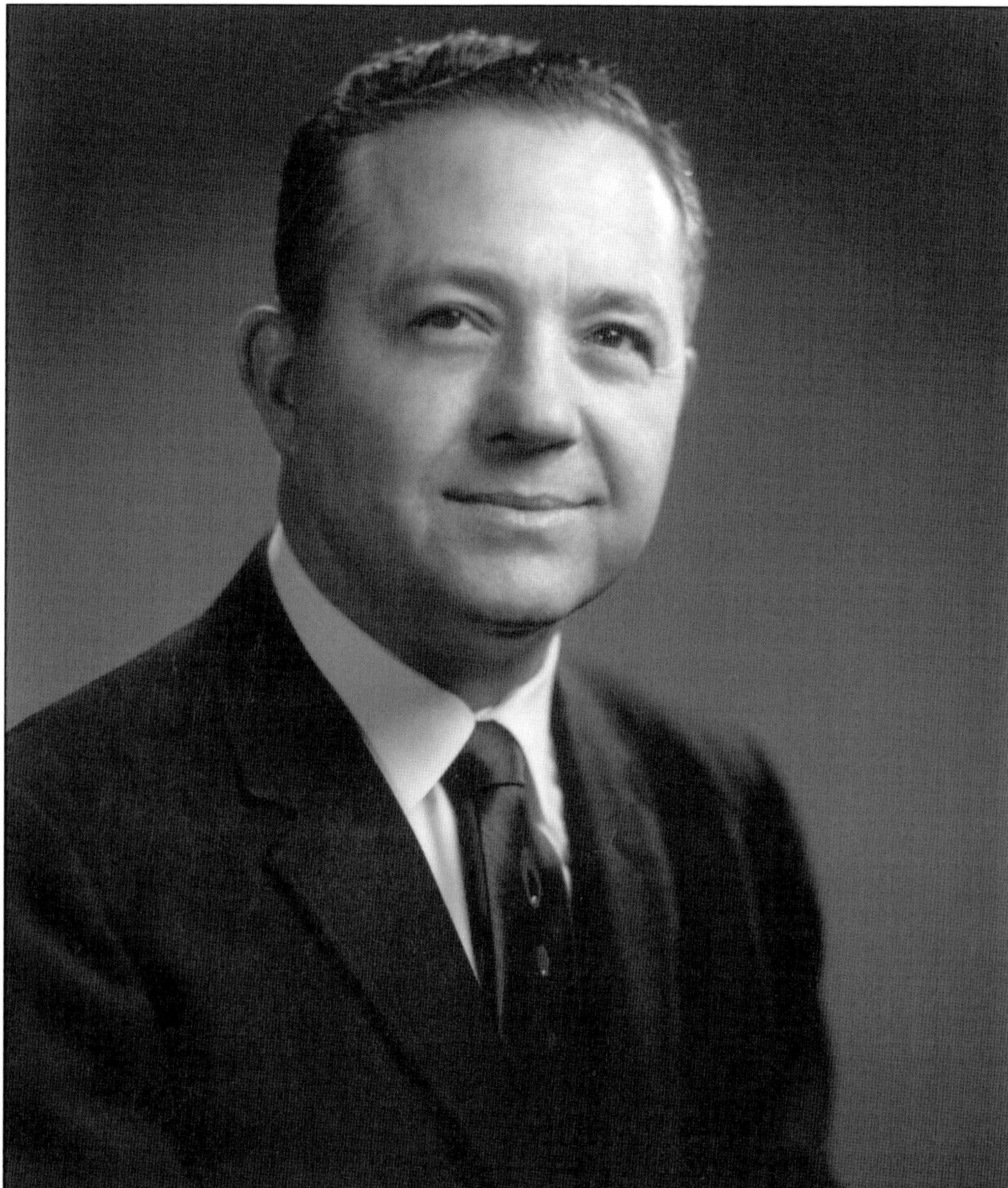

Howard Nicholson, manager of the Memphis branch of Paramount Pictures, and his wife, Rosemary, operated the 51 Twin Drive-In Theater on Highway 51. They sometimes had trouble getting first-run movies and had to sue for the right to show *Walking Tall* in 1973. Many film distributors favored Memphis theaters by withholding movies for about a month from any theater deemed competitive, including Millington's, despite its location 17 miles from Memphis. Long before the Nicholsons opened the drive-in, Martin Williams operated an open-air theater in the heart of old Millington, across from the Strand Theater, which featured silent movies before talkies began in the 1920s. Emery "Dinty" Moore operated a theater with a balcony, also in town, but of course, trips to Hansonhurst were all the more enticing for its upstairs theater above the icehouse. Millington's Strand Theater first aired movies before its reincarnation in the 1980s as a venue for gospel and country music shows.

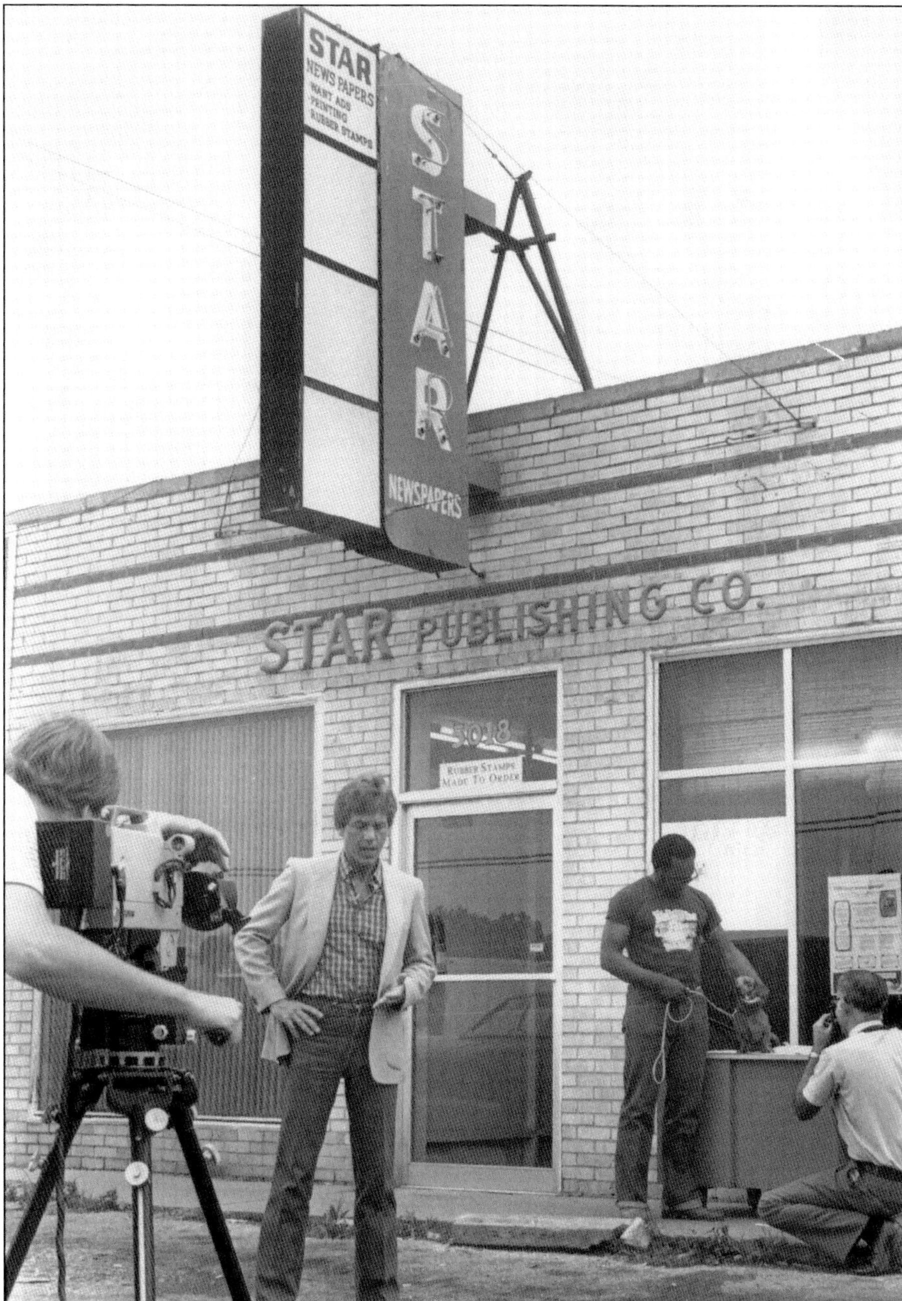

Star Publishing Company is best known for its newspaper, the *Millington Star*, but it also provided printing services and sold rubber stamps. The company not only reported local, regional, and national news, it strove to maintain a good relationship with an actively involved public. One way it managed this was by sponsoring contests. Shown here, second from right, is the winner of the ugliest dog contest with his pooch. The first newspaper in Millington was operated by Ella Ward Allen Bullfin, who met the daily trains to collect the latest gossip for printing. (Courtesy *Millington Star*.)

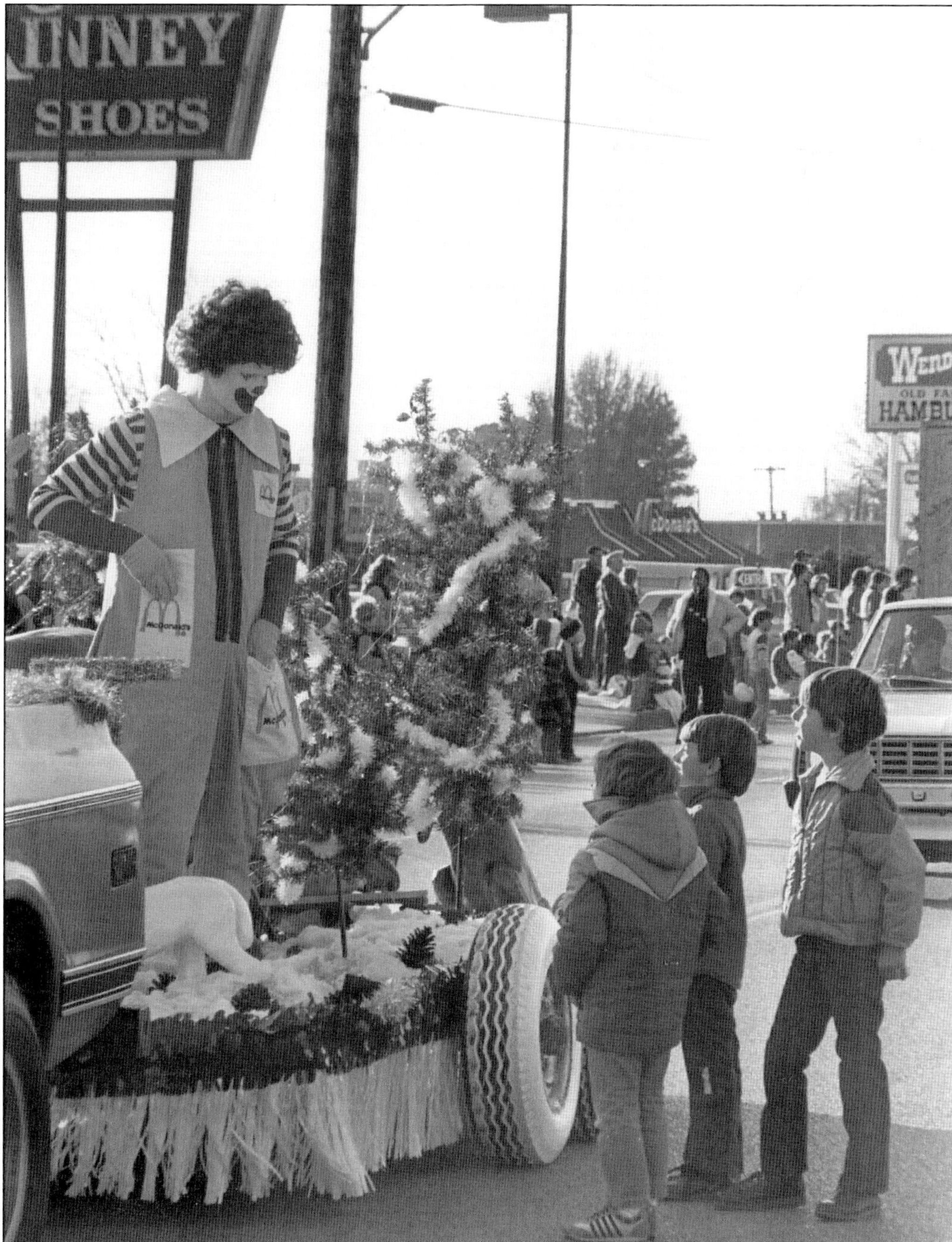

These little boys wait for goodies as Ronald McDonald reaches into his pocket from his perch on a float during the annual Christmas parade down Navy Road. Public celebrations of this holiday have been a tradition since the town's early days. In the 1920s, a town Christmas tree served as an invitation to Santa, who brought fruit rather than toys—a special treat during the winter when fresh fruit was hard to find. In later days, Santa heard children's holiday wishes from his seasonal cottage at the telephone company. (Courtesy *Millington Star.*)

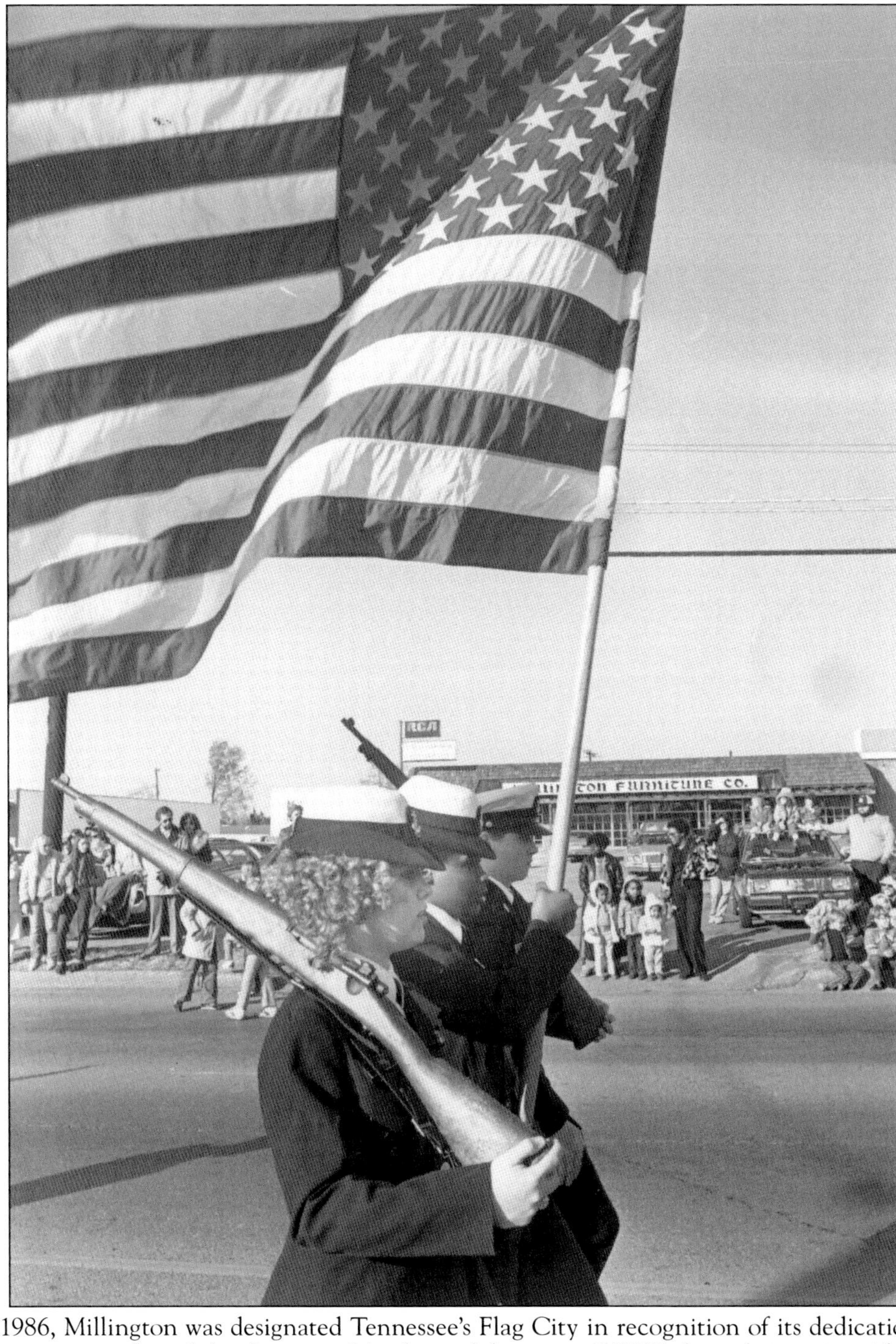

In 1986, Millington was designated Tennessee's Flag City in recognition of its dedication to economic growth and development. Members of the junior ROTC show their civic pride and patriotism as they carry the national banner past Millington Furniture Company. Annual Flag Day celebrations have grown over the years and continue to draw large crowds for entertainment, food, and fireworks celebrating the freedom that the flag represents and which so many past and current Millington residents fought to defend. (Courtesy *Millington Star*.)

Bibliography

Chamberlin, Shirley Sigler. *History of Cuba, Tennessee*. Millington, TN: Shirley Chamberlin, 1984.

Howard, W. S. *Old Timers Restaurant: A collection of rare old photographs depicting bygone days of Millington and the surrounding area*. n.p., n.d.

McCalla, Jon P., Penny Saucier Glover, Ruth Blakey Billingsley, Louise Wooldridge Rhodes, eds. *An Illustrated History of the People and Towns of Northeast Shelby County and South Central Tipton County*. Millington, TN: Historic Archives of Rosemark and Environs, 2010.

Osteen, Faye Ellis. *Millington: The First Hundred Years*. Southaven, MS: King's Press, 2002.

Van West, Carroll, ed. *The Tennessee Encyclopedia of History and Culture*. Nashville, TN: Rutledge Hill Press, 1998.

Williams, Thelma Sigler. *The Early History of Millington*. n.p., n.d.